Dark Land, Dark Skies

This book is dedicated to my wonderful and ever loving wife

Dena Griffiths

May there always be starlight on our path

Dark Land, Dark Skies

The Mabinogion in the Night Sky

Martin Griffiths

Seren is the book imprint of
Poetry Wales Press Ltd,
4 Derwen Road, Bridgend, Wales, CF31 1LH

www.serenbooks.com
facebook.com/SerenBooks
Twitter: @SerenBooks

© Martin Griffiths, 2017
Reprinted 2021, 2024

The rights of the above mentioned to be identified as
the authors of this work have been asserted in accordance
with the Copyright, Design and Patents Act.

ISBN: 978-1-78172-653-2

A CIP record for this title is available from the British Library.

All rights reserved. No part of this publication may be reproduced,
stored in a retrieval system, or transmitted at any time or by any means,
electronic, mechanical, photocopying, recording or otherwise without
the prior permission of the copyright holder.

The publisher acknowledges the financial assistance of the Books Council of Wales.

Cover photograph: © Martin Griffiths

Constellation charts: © 2005 Andrew L. Johnson

Printed by 4Edge Ltd, Hockley.

Contents

Foreword	7
One – The Land and Sky	11
Two – The Welsh Autumn Sky	41
Andromeda, Aquarius, Aries, Capricorn, Cetus, Pegasus, Perseus,	
Pisces Austrinus, Pisces and Triangulum	
Three – The Welsh Winter Sky	77
Auriga, Cancer, Canis Major, Canis Minor, Eridanus, Gemini,	
Lepus, Lynx, Monoceros, Orion, Puppis and Taurus	
Four – The Welsh Spring Sky	127
Bootes, Canes Venatici, Coma Berenices, Corona Borealis, Corvus,	
Crater, Hydra, Leo and Virgo	
Five – The Welsh Summer Sky	161
Aquila, Cygnus, Delphinus, Libra, Lyra, Ophiuchus, Sagitta,	
Sagittarius, Scorpius, Scutum, Serpens and Vulpecula	
Six – The Welsh Circumpolar sky	217
Camelopardalis, Cassiopeia, Cepheus, Draco, Ursa Major	
and Ursa Minor	
Conclusion	242
Glossary	244
Bibliography	257
Index	258
About the Author	266

Foreword

One cannot escape the majesty of the heavens, the overwhelming perception of being a very small part of such a monumental edifice as the starry sky above us. Yet at the same time, one cannot help but feel at one with it. Such is the magnitude and wonder of our universal home that it comes as no surprise that the heavens have been studied from the time that man first walked the Earth. Every ancient civilization looked to the stars, grouped them into constellations and imbued them with a narrative hoping that the wisdom seen in the night sky would have a marked effect upon the course of life they could lead.

The ancient peoples of Britain, and especially the Cymru or the tribes that eventually would live in the land of Wales, also had their own cultural affinity for the sky. Many of the tales they told were shared in oral traditions that have been lost over the centuries. Others were recorded in post-Roman Britain by bards who kept the traditions alive.

In the immortal words of Monty Python: "what have the Romans ever done for us"? Well, they left us an invaluable system of writing and Roman script, so ancient tales began to be recorded by those who could read and write after the Roman system of education. This script was used to create the first poems and stories about Wales that drew upon some earlier oral traditions and tales. The oldest collected tales are from sources in thirteenth century Wales and probably date back to the fifth and sixth centuries, the time of the poets Taliesin and Aneirin. Both feature the lives of the princes of an extended kingdom of Wales: Aneirin was born in Edinburgh and wrote his classic tale of battle and loss *Y Gododdin* after the Anglo-Saxon invasion of eastern England. He is the first to mention some of the deeds of King Arthur. Conversely, Taliesin probably moved from court to court as his works in the *Book of Taliesin* praise king Urien of Hen Ogledd (now northern England and southern Scotland) and Brochfael Ysgithrog of Powys amongst others. This shows that Welsh as a language was highly influential in Britain at the time, and was probably the *lingua franca* of most of the country. So, tales of the great deeds and how they fitted into the sky would be commonplace in Celtic, early English and Pictish culture due to this literary influence.

The majority of old Welsh tales and poetry are known from a few volumes, *Hanes Taliesin*, the *Black Book of Carmarthen*, the *Red Book of Hergest* and the *White*

Book of Rhydderch. Collectively some of these tales, poems and insights became the *Mabinogion*, the folk tales of Wales which were translated into English and published by Lady Charlotte Guest, albeit in a rather bowdlerized Victorian fashion, as a way to bring these tales to the masses that flooded into Wales during the eighteenth and nineteenth centuries to work the iron and coal that put the country at the forefront of the industrial revolution. It also raised the language and culture of Wales to a status probably not enjoyed previously and gave Welsh literature, folklore and storytelling a similar status to classical Greek and Roman myths and poetry.

Today the *Mabinogion* and its related texts are a great resource of academic scholarship and argument. But that is not what I intend to explore here. Instead I will take some of these ancient tales and marry them to the constellations that they pre-figure in the sky or at least are associated with in legends and tales across the Celtic systems of Wales, Ireland and Scotland, including that of some English folk tales that have an origin in Celtic myths. I will do this to show that there is a rich culture of sky lore that ties events, stories and meanings together in a way that is fairly unique to Wales, and which is as important as the classical Greek myths that are usually used to illustrate the night sky.

Also of importance to this book is the fact that the night sky is threatened by increasing light pollution and that the education system that disseminates such myths, in English, Welsh or science classes, is also threatened. Many people in urban areas have never encountered the Milky Way or seen a truly dark sky. Additionally, they have probably not been introduced to the night sky or its store of treasures in a way that ties the wonders of our modern understanding to the cultural roots of the past. This book is intended to redress that balance, not only to show some of the beauty of the night sky, but to marry that beauty to the ancient Celtic landscape and tales of Wales.

The number of dark sky parks and dark sky reserves is now increasing due to the coordinated efforts of the International Dark Sky Association, the Commission for Dark Skies in the UK and other interested bodies globally. The scourge of light pollution and the carbon footprint necessary to generate the power to light up the night sky needlessly is now being recognized by smaller groups and public interest bodies such as local councils and national parks.

Wales is fortunate to have two International Dark Sky Reserves: Brecon Beacons National Park and Snowdonia National Park, areas where the dark night sky is preserved for future generations. These beautiful landscapes and their pristine skies are joined by the Elan Valley Dark Sky Park in mid-Wales and by ongoing work toward ensuring dark sky communities in Anglesey and the Pembroke Coast National Park. Wales may truly be called the first Dark Sky Nation, a fact of which we can all be proud as the heritage of both land and sky are safeguarded for future generations.

Capitalizing on this status it is possible to build and utilize a small observatory for public education and to enhance the experience of the night sky for everyone. The Brecon Beacons International Dark Sky Reserve has such an observatory and a teaching classroom at the National Park Visitor Centre, which has been used extensively for training and for public events since it opened in 2014.

Using such an observatory can be a wonderful experience, especially for urban-based astronomers who don't have access to very dark skies. Even those who may have portable equipment and have taken advantage of the dark sky status of the National Park enabling them to enhance their viewing will find the facilities at the observatory will allow access to the wonders of the night sky that they cannot reach from light-polluted areas.

In the case of the Brecon Beacons Observatory (BBO), its 30cm f5 reflector on a driven EQ6 mount thrilled over a thousand visitors in its first year of operation. Fitted with a piggybacked 120mm refractor for DSLR imaging or just visual observing and an Atik 314L CCD camera for imaging of objects, this small observatory has added to the experience of tourists and local astronomical societies within the national park and in south Wales generally. The BBO also has the advantage of a classroom at the visitor centre to enable education throughout the year. It is a place where the public can receive astronomy presentations and enjoy the warmth and conviviality of hot drinks on tap!

George Borrow, the nineteenth century Victorian gentleman traveler wrote in his book *Wild Wales* of the brooding, dark landscapes he encountered as he traversed the country. Those landscapes still exist in the heights of the Snowdonia mountains, the Cambrian range, the wine red hills of the Brecon Beacons, the shaded uplands of Mynydd Preseli and in the moorland landscapes of mid Wales. The mysterious waters of Wales' lakes, mountain tarns, rivers and seas are at the heart of many tales that tie semi-mythical figures who trod the land with the starry patterns of the sky. That is the landscape and skyscape that I wish to bring to life and share with generations to come: the spectacle, wonder, curiosity and pride aroused in me from these old tales that I first learned as a boy. I hope that I have done them justice.

As so many of these old tales have been passed down orally, many aspects of them have changed over the years. Some of the associations between land, tale and sky are just discernable, whilst others are obvious in their placement with a particular constellation. Others share common themes across many cultures, and tales were probably shared among peoples, evolving and dispersing as their cultures fractured after the Roman withdrawal from Britain and greater Europe. It is not my intention to gerrymander tales into particular groups, but to make those possible connections plain and tell not just the Welsh side of the story but to include the classical interpretations too.

I hope that the reader enjoys this journey through the Welsh mythological landscape. I also hope that knowing some of these tales and their heavenly associations will bring a new interest to the night sky and any stargazing experience they will have.

Martin Griffiths BA BSc MSc FRAS FHEA
Brecon Beacons Observatory

CHAPTER ONE
The Land and Sky

To live in Wales is to be conscious
At dusk of the spilled blood
That went into the making of the wild sky,
Dyeing the immaculate rivers
In all their courses.
It is to be aware,
Above the noisy tractor
And hum of the machine
Of strife in the strung woods,
Vibrant with sped arrows.
You cannot live in the present,
At least not in Wales.
There is the language for instance,
The soft consonants
Strange to the ear.
There are cries in the dark at night
As owls answer the moon,
And thick ambush of shadows,
Hushed at the fields' corners.
There is no present in Wales,
And no future;
There is only the past,
Brittle with relics,
Wind-bitten towers and castles
With sham ghosts;
Mouldering quarries and mines;
And an impotent people,
Sick with inbreeding,
Worrying the carcase of an old song.

– R S Thomas

Chapter One – The Land and Sky

Celtic peoples were living in Britain long before the arrival of the Saxons, Angles and Jutes from Friesland and Jutland whose languages would eventually develop into English. No doubt they had a varied selection of tales and stories that were significant to them. Wales, Ireland, Scotland and England (Angle-land) were subject to a series of invasions in the first millennium AD that are recorded, not always accurately, by historians such as Bede and Geoffrey of Monmouth, though in all probability there were centuries of trade and interchange with the tribes of the northern European continent. Modern genetic studies reveal that there is very little difference between the so called Celts and the Anglo-Saxon people.

Wales as a country however, always seemed a strange land to these visitors. Despite Celtic people populating the British Isles after the last ice age, between 12,000 and 15,000 years ago, it was the Celts and not the Anglo-Saxons who came to be called 'strangers' or foreigners. The modern English word for the descendants of the Cymru, just one of these Celtic peoples, is *Welsh*. The name Wales, comes from the Norse word *wealh*, meaning foreigner, stranger, or in this case, Celt. Its plural *wealas* is the direct ancestor of the modern name: Wales. To these ancient peoples from the lower countries of Europe, a landscape with wild hills, extreme weather and belligerent tribes would have made an impact that caused them to think of this land as shadowy, menacing and mysterious. What culture could possibly emerge from such a background?

Today, Wales may be a small nation, but it has a long and rich sense of history, a language and culture that is found nowhere else, a sense of pride in the past and a strong perception of its difference, all married to an incredible diversity of landscape. From some of the highest mountains in the UK in the north of Wales, through the rural green heartland of Powys and the highest mountains in southern Britain in the Brecon Beacons, to the farmland of the Vale of Glamorgan and the rugged coastline of Pembrokeshire with its dramatic Preseli and Carningli Hills, Wales offers so much to its inhabitants and the millions of tourists who visit each year.

The waterfalls, the brooding hills, the wilds of its moorland and the dramatic sweep of its high mountains combine with a wonderful coastline where sandy beaches, tiny bays and rocky cliffs merge to produce some of the most beautiful scenery to be found anywhere on earth – and all of it can be reached in a few moments or in a few hours drive. The diversity of the Welsh landscape is only matched by the diversity of its weather, yet despite its high rainfall, the crystal clear air left after a summer shower or the cold, deep black of its winter skies draws people to this enchanted land where history, culture, poetry and stories all mingle together to produce one of the most singular cultures in Europe. This ancient culture has left its mark on the landscape, its people and upon the sky.

The mild maritime air that blankets Wales makes night time observing in any season a joy, whilst the wonderful landscapes and ancient ruins scattered in profusion across the principality offer the daytime visitor a sense of the continuity between people, land and sky that is being lost in our modern world. Wherever you are, the night sky is one of the wonders of nature; a profound sense of reverence and awe grips anyone who looks up and truly contemplates the scene of thousands of stars resolutely pursuing their course across space; the grandeur of the Milky Way on a summer's evening or the scintillating beacons of the winter night. Currently Wales offers a matchless opportunity for astronomy through its two International Dark Sky Reserves and its International Dark Sky Park.

Looking up at the night sky from these locations we recognize the classical groups of the constellations. Although astronomers in the west generally follow the Greco-Roman outlines and mythologies, in reality the groupings are very old, varying in composition and grouping across some cultures. The constellations we currently recognize were first recorded by Mesopotamian astronomers in the sixth millennium BCE; more a testament to the fact that these people invented writing rather than inventing the constellations. In the Lascaux caves in the Dordogne of France there are fantastic cave paintings drawn by the Cro-Magnon people 30,000 years ago. Amongst all the pictures of extinct animals is the figure of a bull, but it is an interesting figure.

In all of astrology and astronomy the bull Taurus has always been drawn as the front half of the animal with long horns and big shoulders. This is exactly how this bull in Lascaux is drawn. How do we know that it is a constellation? Whoever drew the figure took the time to dip his finger in charcoal and drew six stars on the shoulder as a likeness of the Pleiades, so this deliberate act draws attention to the fact that at least some constellations, especially those of the zodiac, were known to ancient peoples and passed along in oral traditions. Why paint Taurus? With modern computing and science, we have discovered that the Earth wobbles slightly on its axis; 30,000 years ago the position of the Sun entering this constellation brought on northern summer as the Sun (apparently) moved northward. The fact that these people knew of this motion shows that they had been sky watchers for a long time.

The Lascaux painting is a visual representation of something that could only be understood by those listening to a tale; how much more did they know about the sky? It is a pity that we will never have an answer to that question.

When it comes to the patterns of the constellations, ancient Europeans, including the Celts followed the groups as already formed by older civilizations. The Celts, who were a disparate series of tribal cultures spread across northern Europe in the first millennium BCE, did not have written records but oral traditions and it is only their contact with the Roman world that leaves us anything of their culture, beliefs and society. The tales they may have told each other and the bardic traditions of the Gauls, the Irish, British and Pictish tribes may be a left-over of myths that originated in the heroes that populated the sky but it is hard to be absolutely positive about this.

Nevertheless, there are enough associations between ancient tales and skylore to allow us to assess the stories and fit them to some of the constellations. Roman writers such as Julius Caesar and Livy point to the central importance of religion and rites in Celtic life, though exactly what the Celts believed and how each god or goddess dominated any particular rite is now obscured by ancient bias, re-interpretation through archaeology and the rise of new age religions and paganism. That they had animistic beliefs is not contended; many cultures of the same period did and so the association of spirituality with trees, streams, pools, stones, plants, mountains and sky in Welsh Celtic culture is probably fairly evident even if we have no direct and absolute proof. Just as the Romantic poets felt an attachment to the natural world, so did the ancient Celt who probably saw the landscape resonating with a significance that is difficult to recapture.

The only people historically who had first hand information on the Celtic tribes of Wales were the Romans, who left all sorts of information on the practices of human sacrifice by the Druids, whilst at the same time admitting that these priests functioned as physicians, judges and advisors. They relate that the Druids held the power of knowing which gods to placate, when to plant, when to hunt, when to harvest and when to call for religious rites. None of this could be done without at least a rudimentary knowledge of the movement of the sky through the seasons and this would no doubt have been accompanied by various stories of what each constellation and planet meant in rhyme and story, to be passed on in oral tradition from priest to initiate.

The Romans found them to be a powerful enemy and ultimately destroyed the main Druid strongholds in northern Wales and Anglesey. However, perhaps enough of their stories survived to become part of the Celtic lore that resonates down to our day via the ancient bardic tradition. At least one such ancient tradition survives: that of 'kissing under the mistletoe' at Christmas, which is meant to promote fertility and banish barrenness in the new year, so who knows what else may survive?

The tales recorded in the ancient books of Wales tell of chivalry and romance, quest and hardship, mystery and magic, triumph and sorrow, epic battles and long years of plenty; they contain the origins of Arthurian legend and speak of the wisdom of animals and the natural world around us that is there to be observed if one takes the time to commune with it. These tales have many parallels with, and possible origins from, the vibrant stories the Celts attributed to the stars of the night sky.

The dark land and the dark sky are epitomised in many Welsh tales. The interplay of humanity, landscape and the natural world around us is evident in the inclusion of hills and vales, mountain passes, coastal areas, place names that we recognize in this modern day; all matched to the flora and fauna of sea, land and sky. Tales such as Culhwch and Olwen, Pwll, Lord of Dyfed, Branwen, the Daughter of Llyr and others show remarkable relationships between the world of men and the secret world of nature. It just takes a prepared heart and mind to enjoy the *hiraeth* of such stories, to suspend your disbelief and position oneself in this mysterious world, to enjoy a harmonious rapport with saga, stars and society.

Observing the Sky

Despite their longevity and human association, constellations can be confusing and indistinct if you are exploring them for the first time; how is it possible to connect all those tiny lights in the sky into recognizable patterns? Some of these patterns may be recognizable, such as Ursa Major or Orion, but others may be lost in the confusion of different colours, brightness and lack of structure. It is probably best to use the constellations one recognizes first to then find other, fainter groups. It is important to realize straight away that one will not be able to see all the constellations of the sky at once, yet this works to the advantage of the stargazer that is starting in their quest to understand these celestial groups.

A planisphere is a useful piece of equipment that will help you to find the groups visible each night of the year. They are available in good bookstores, or can be found online. Modern technology is also useful as here are apps for iPad, iPhone and Android so that one can use a smartphone or tablet to explore the sky. The charts and descriptions in this book can be used to plan what you wish to observe. Within the constellations themselves, there are suggested deep sky objects that comprise a collection of the best objects to see, but is not a complete or extensive range. Additional research and application with telescopes will bring out many more.

For the sake of simplicity, the stars on both seasonal and constellation charts have been connected together with 'imaginary' lines that give the figure a fuller appearance. Naturally, you will have to imagine these lines connecting the stars when you look at the sky, but this is not as hard as it sounds. In fact, using such lines, you will quickly discern the patterns of the constellations and begin to think in terms of patterns rather than in disconnected points of light. The ability to recognise patterns is the key to understanding the night sky. Thankfully, you don't have to impose order on the sky as it has already been done for the observer – we are recognizing patterns set down in antiquity.

One caveat if one is using software or apps on a phone or iPad; some planetarium programmes overlay the constellation stars with the stylised figures they are meant to represent, which is very nice but can sometimes add to the confusion first experienced as the figure bears no relation to the star positions! Some of the constellations are created from faint stars with no discernible pattern to the group. When one looks at the sky, there is no substitute for experience and familiarity, and that only comes with practise and a little dedication. Recognizing constellations is a skill that modern man has not learned; therefore, you will acquire a talent that very few possess.

Practise makes perfect. It is essential to go out and find the constellations whenever the sky is clear, as doing so will refine your abilities and enhance what you have already learned. Additionally, the constellations move during the night or over the months; circumpolar ones wheel around into different positions and their configuration may be lost if one is used to finding them at one particular location in the sky. Some observers also tie the positions of certain constellations to a landmark they can recognise, but forget that the sky is dynamic and the constellation may not stay tied to that arrangement of Earth and sky for long. Nevertheless, enjoy the knowledge and the new things learned. Put together with everything else one can learn from the sky and associated disciplines, your interest will become a complete education.

Constellations can be separated into convenient groups that broadly follow the seasons. Thus there will be groups that correspond to autumn, winter, spring and summer with the constellations of the northern circumpolar sky forming a permanent group that is always visible to those living above 40° north latitude. To gain a foothold it is important to recognize the major stars of the seasons and then to pick out a single constellation from each season to act as a guide to the others.

Before one goes rushing out under the night air to discover all the heavens have to offer, it is wise to consider a few questions. Are you going to become just a naked eye observer? Will you be able to remember the constellations as they assume different positions in the sky at different times of year? What sort of observing aids will you use to explore the sky? These and many other things must be considered to get the most from your observing experience.

Remember that no matter how rich or poor one is, we all have access to the same universe. The constellations are the same for everyone, but knowledge of them is not. Therefore, even using just the naked eye to explore the heavens is important, as is the ability to remember the stories that go with the constellations from many different cultures. Traditional societies value the ability of storytellers who can pass on an ancient oral custom and even in the developed nations, the ability to tell stories, relate myths and know the sky intimately gives one a specialized knowledge open to very few. Knowledge of the night sky is a specialism to be proud of, but the point of such education is simple: *learn and pass this knowledge on.*

Common Classifications

In the seventeenth century the astronomer Johann Bayer (1572-1625) introduced the cataloguing of the brightness of each star within a constellation. To avoid confusion, Bayer thought it appropriate to arrange the stars in any particular group by means of the Greek alphabet. Thus the brightest star of a constellation would be called alpha, the next brightest would be known as beta, then gamma, delta, epsilon... and so forth. The Greek alphabet appears below (each column should be read top to bottom and left to right):

α Alpha	ζ Zeta	λ Lambda	π Pi	φ Phi
β Beta	η Eta	μ Mu	ρ Rho	χ Chi
γ Gamma	θ Theta	ν Nu	σ Sigma	ψ Psi
δ Delta	ι Iota	ξ Xi	τ Tau	ω Omega
ε Epsilon	κ Kappa	ο Omicron	υ Upsilon	

The arrangement worked brilliantly, and has been the accepted way of designating the stars in a constellation ever since. When someone talks about the star α Orionis or β Cassiopeia, we know that they are talking of the brightest and second brightest star in those respective constellations. Bayer made very few mistakes with this general classification, and its success is a testament both to its simplicity and the genius of this underrated astronomer.

Other methods of stellar classification were developed by assigning the stars certain numbers, the number depending on their right ascension. This was undertaken by John Flamsteed (1646-1719), the first Astronomer Royal. His system was to give the number one to the star with the lowest right ascension in a constellation, then proceeding eastward and assigning numbers 2, 3, 4, 5 and so on in that order. Today his numbers are not generally used for the brighter stars of a constellation, but are widely used for the fainter stars. Take a glance at an acceptable star atlas and the stars will be numbered using this system.

This system, and the charts of 30,000 stars that become the *Atlas Coelestis Brittanica* was Flamsteed's major contribution to observational astronomy. After Flamsteed's death, he was succeeded as Astronomer Royal by Edmund Halley, who computed the orbit of the comet that bears his name. Just to reveal that not everything runs smoothly in science, Flamsteed became an avowed enemy of Halley, and upon his death Flamsteed's widow stripped the Greenwich observatory of all her husband's instruments, leaving Halley to finance the observatory himself. Despite this setback he completed Flamsteed's work to produce the most comprehensive star atlas of the northern sky, which became the workhorse of British observational astronomy for the next hundred years.

However, it soon became necessary to impose accurate observational constraints on the positions of stars – a process known as Astrometry, especially after naval latitude and longitude were introduced. The latitude and longitude of the sky is necessary to give accurate positions of stars and follows the pattern of naval navigation in that it uses degrees, minutes and seconds.

Right Ascension & Declination

The systems of right ascension and declination are not necessary for one to learn in order to find the stars of the constellations. However, these coordinate systems will become essential once you are familiar with the sky and wish to find objects from their given positions in a star atlas or magazine. This aspect of celestial navigation is not as complex as the terms sound, certainly, if you can read a map, then you can find your way around the sky.

Right ascension is the celestial equivalent to lines of longitude on the Earth, only instead of degrees in this case, we use hours, minutes and seconds that correspond to the hours recorded here on Earth. As Greenwich Mean Time is used as the standard for time on the Earth's surface, then a point in the sky has to become the 'zero' hour for celestial longitude too. This point of 00 hours of longitude or 'right ascension' (RA) is the point where the Sun crossed the celestial equator on March 21st each year and can be found as the 'First point of Aries' (even though confusingly the point is now in Pisces). The whole sky of 360° is divided into 24 hours – the same time taken for one revolution of the planet. Each hour of RA corresponds to 15° of longitude and each hour is divided into 60 minutes, and 60 seconds, usually presented in shorthand as ' for minutes and " for seconds. They are known as minutes and seconds of arc (as the sky is a 'circle') or arc-minutes and arc-seconds.

When reading the coordinates off a star map, always quote the right ascension first, in hours and minutes, then read off the declination. The declination corresponds to the lines of latitude on the Earth. There is a celestial equator, but no tropics. Anything north of the equator becomes a + declination, whilst everything south of the equator becomes a − declination. Declination is measured in degrees, not hours and minutes, although degrees are divided into minutes and seconds also, giving a fine adjustment and increasing accuracy. If you wish to consult an ephemeris to find an object in the sky, always remember that the RA will come first given by the declination; e.g. RA 12h 34m 14.5s + 34.53.21.3. The RA can sometimes be shortened to 12.34.14.5 where the sequence of numbers corresponds to hours minutes and seconds. The same is true for the declination.

Stellar Classification

Most people are completely unaware that stars have different colours. Colour perception is a personal thing and just a little training will enable one to identify these colours quite easily. This colour difference becomes extremely important as colour is related to the temperature.

Harvard University was the workshop for stellar grouping. At the beginning of the twentieth century, the eminent astronomer Edward Pickering brought together a fine group of young ladies to do the work of assisting him in his quest to catalogue stars by spectral appearance. This was an unusual step when astronomy was generally considered a gentleman's pursuit. As a result, the group become affectionately known as 'Pickering's Harem', but these young ladies proved to be a valuable resource, several of them, such as Henrietta Leavitt, Cecilia Payne-Gaposhkin and Annie Jump Cannon, made discoveries of literally cosmic importance.

Recognizing the importance of colour as a function of temperature, the Harvard ladies first thought to catalogue the stars by using the English alphabet ABCD… etc. However, the constraints placed upon the colours and spectral characteristics of stars led to the representation that is used today as many of the original classifications turned out to be false leads or were repetitions of other stars. This classification by colour revealed that the hotter the star, the more blue it was; the cooler the star, the redder it was. This reflects the electromagnetic spectrum in which the component of visible light that we see

Spectral Type	Colour	Temperature (k)*	Spectral Features
O	Indigo	28,000-50,000	Ionized helium, especially helium
B	Pale Blue	10,000-28,000	Helium, some hydrogen
A	White	7,500-10,000	Strong hydrogen, some ionized metals **
F	Pale Yellow	6,000-7,500	Hydrogen and ionized metals such as calcium and iron
G	Deep Yellow	5,000-6,000	Both materials and ionized metals, especially ionized calcium
K	Orange	3,500-5,000	Metals
M	Red	2,500-3,500	Strong titanium oxide and some calcium

* To convert approximately to Fahrenheit, multiply by 9/5.
** Astronomers regard elements heavier than helium as metals.

between 390 and 700 nm (Nanometer or nm = 1 billionth of a metre or 1 x 10^{-9} m) follows the grouping of colours familiar to us from the rainbow. The shorter (violet) end of the spectrum is at 400 nm and the longer wavelengths (red) are close to 700nm.

The grouping that the Harvard computers eventually settled on as representative of all stellar spectra, is simply this: O B A F G K M in descending order of temperature. The classes can be remembered by the simple mnemonic "Oh Be A Fine Girl Kiss Me". These are examined below:

Type O stars are blue-white and extremely hot, typically around 25,000 K and higher. These very massive stars are very luminous, and are also the most short-lived. Their spectra show lines from ionized helium, nitrogen, and oxygen. A typical example of this class is Iota Orionis.

Type B stars are blue-white and very hot with temperatures of around 20,000 K. Stars of this type are generally massive and quite luminous. Their spectra display strong helium lines. Rigel, Spica and Regulus are good examples of this class. Compare Rigel and Regulus as an exercise.

Type A stars are white, with temperatures of around 10,000 K. Their luminosities are usually about 50 to 100 times that of the Sun. At these temperatures no helium lines are present, but strong hydrogen lines appear in the spectra. Sirius, brightest star in the night sky is of this class as are Vega and Altair, part of the summer triangle.

Type F stars are yellow or yellow-white, with temperatures of about 7,000 K. Their spectra show weaker hydrogen lines, but strong calcium lines. In some of the later classes the spectra of metals begin to appear. The winter star Procyon in Canis Minor is of this class.

Type G stars are yellow, with temperatures of about 6,000 K. Their spectra show weaker hydrogen lines, but stronger lines of many metals. The Sun is a typical G type star.

Type K stars are orange, with temperatures of about 4,000 to 4,700 K. They have faint hydrogen lines, strong metal lines, and hydrocarbon bands in their spectra. Aldebaran in Taurus and Arcturus in Bootes are good examples.

Type M stars are red, with temperatures of about 2,500 to 3,000 K. They have many strong metallic lines and wide titanium oxide bands and other exotic compounds in their spectra. Betelguese and Antares are spectacular and easily visible examples of this class.

Observing preparation

Most people underestimate the power of the naked eye, and how to use it properly to gain the most from astronomy. Simply put, the eye is the most sophisticated optical device known to man. Telescopes and binoculars are supplements to this wonder of the human body, and many people are surprised at what the eye alone can discern. The naked eye can discern details as small as one arc-minute on an astronomical body, a size smaller than some of the Moon's largest craters. On a dark night, the iris, the diaphragm controlling the amount of light entering the eye, is fully open and the pupil measures on average about 5mm. At that point, the eye is obtaining the maximum input from the surroundings and can even discriminate between subtle areas of light and shade.

When you go out into the darkness of an evening, take a little time to allow the pupil to open fully. This technique is known as Dark Adaption. The process can take about 5 to 10 minutes, but to be fully dark-adapted takes up to 30 minutes. During this period of time, avoid lights of any kind, do not switch on your torch or stare at nearby street lighting or the adaptation is lost.

Once this has been accomplished, you can begin observing, as your eyes are now a little bit more sensitive to light than they ordinarily are during the day. However, some astronomers have noted that faint objects seem to be a little brighter if they are seen 'out of the corner of the eye', as it were, and this phenomenon is an important tool of the observer. The process of seeing objects in greater detail simply by not looking directly at them is called Averted Vision.

To enable you to maintain your night vision, it is best to examine any star charts or atlas by means of a torch with a red beam. This red light will not interfere with a dark-adapted eye, and is comfortable and easy to see such charts by. When observing, make sure that your comfort is the paramount consideration, so always wrap up warm, have a hot drink handy and take a break every hour or so if you intend to observe all night. Additionally, if you can stand on a raised board whilst observing, then the heat of your body will not be sapped through your feet, leaving you cold and miserable. This is simple common sense, yet many observing sessions have been ruined by the lack of such preparation.

To see deep sky objects and faint stars at their best, it is best to avoid times of the month when the moon is shining brightly. Whilst the moon is a lovely romantic object shimmering with a silvery light, looking wonderful in a cloudless sky, it is less than romantic to astronomers and photographers interested in digging out remote or obscure clusters, nebulae and galaxies. During an average month there is at least a two-week moon free period, when deep sky objects can be seen at their best. In addition, only observe when the air is fairly clear and the atmosphere, or seeing conditions remain quite steady. High, hazy cloud does rather spoil the view.

Thus, seeing the sky at its best takes a little patience. Even the most perfect summer day can affect the observing conditions at night as the heated atmosphere gives rise to tremulous effects which then have an impact on the astronomical 'seeing', a term that describes the properties of the atmosphere. Turbulence from heat rising, and humidity all detract from the perfection of crisp point-like stars.

If you are going to use a telescope or binoculars, it is imperative to find a dark sky site. This is a site where no street lights or the pervasive glow of street lights can be seen, and the sky remains dark right down to the horizon without interruption from any light sources through a 360 degree circle. Granted such sites may have to be found well outside your district or area, especially for city bound astronomers, but it is worth the effort to go out of your way to obtain fine observational results. In Wales we have the Brecon Beacons National Park International Dark Sky Reserve, the Snowdonia National Park International Dark Sky Reserve and the Elan Valley Dark Sky Park along with the Dark Sky Communities of Anglesey and the Pembroke Coast National Park. We are spoiled for choice!

Choosing and using Binoculars

A pair of binoculars is one of the most underrated pieces of equipment. Most novices think that a telescope is a must, but a good pair of binoculars will reveal much more than a poor telescope. There are so many cheap telescopes on the market, which look appealing but in operation, field of view, stability and colour rendition are quite often poor to the point of being a waste of money. It cannot be stressed enough; invest in a good pair of binoculars rather than a cheap telescope. The image on page 24 shows a selection of binoculars for astronomy.

Compared to a telescope, binoculars have certain advantages. Although they are smaller and have lower magnifications, they are lighter, more portable and easier to use and are less expensive. They give a much wider view than a telescope, thus making objects easier to find. Binoculars also let you use both eyes, providing surer, more natural viewing. In addition, everything seen through them is rendered the correct way up, not the upside-down or backward way that a telescope presents.

It is not possible to recommend the best types of binoculars as optics vary between different manufacturers, but as a general guide it is essential to remember that the larger the aperture, the clearer the view, the less magnification the better. This can be demonstrated by what is known as the aperture index. Binoculars have two numbers set into them near the eyepieces or quoted on the box. These are usually rendered as 10 x 50 or 10 x 25 or another combination of numbers. What this means is that the binoculars have a magnifying power of x10 and the aperture of the main object glass is 50mm. To ascertain the aperture index, simply divide the magnification into the aperture, which with a pair of 10 x 50s is 10 ÷ 50 = 5. A pair of 8 x 40s will have the same index and a pair of 7 x 50s will have an index of 7. The larger then aperture index, the better for stargazing they will be. An index of around 5 is good, anything below that is not really useful.

Many people complain that they cannot see much through binoculars as they cannot keep them still enough or on target long enough due to the movements of the body. Actually one can turn this to your advantage. Evolution has turned man into a predator with our eyes fixed to the front of our head to see and judge the distance of prey. Our eyes once fixed on a target stay on target no matter what the movement of the body. So, if you have an astronomical object in your sights, gently move the binoculars in a rounded motion around the target and the eye will naturally stay on the target regardless of what you are doing.

Similarly, with sighting the object initially, just look at the area of sky with your target in and without moving your head simply bring the binoculars up to your eyes. You should have little difficulty in seeing the magnified image at once. Try both of these methods if you do have difficulty looking for and maintaining a lock on objects. Of course in today's market you also have image stabilized binoculars which enable no shaking of the image no matter what the body movement is (unless you fall over), but such items can be very expensive!

If you are observing with a pair of binoculars, especially if they are the giant sets one can obtain today, you will find after a while that your arms tire and that you develop a pain in the middle of the back caused by bending over backwards to bring the instrument to bear on stars near the zenith. To deal with both these problems, invest in a tripod to steady the binoculars, preferably of a type that has a canted head enabling an observer to get under the tripod to observe the zenith. Even an ordinary tripod is beneficial, as a properly mounted pair of binoculars will show much more detail in hazy objects than an unsteady hand held pair of even the finest binoculars.

Choosing and using a telescope

If you own, or wish to own a small telescope, you must determine your priorities. What do you most want to look at? How dark is your sky? How experienced an observer are you? How much to you want to spend? What storage space do you have, and how much weight do you want to carry? Answering these key questions and familiarizing yourself with what's on the market will enable you to acquire a telescope that will work well and satisfy your observing needs for many years. Plus, you can always trade it in for a larger one!

One of the first considerations is the mount. Most small telescopes do not have very good mounts that are stable and vibration free. Many sold cheaply are simple alt-azimuth mounts with a few screws to secure the scope, which can lead to trouble selecting your object and keeping it in the field of view as one tightens the screws. Even when in use, simply brushing the scope means that it moves off target and acquiring your image again can be a difficult process. If you are purchasing from a supplier, check reviews before selecting a scope, as a good mount is essential to your observing experience.

An equatorial mount is to be preferred as they are a little heavier, have to be set up properly, but once achieved, only one axis needs to move once you are locked on your target. Additionally, the advantage of equatorial mounts is that they can be driven with small motors and they can be used as camera platforms. An alt azimuth mount is only an advantage if one is purchasing a reflecting telescope on a Dobsonian mount. For any other telescope this author considers them to be severely limited. The stargazer must make decisions based on their requirement and budget but there are some excellent quality telescopes on the market with equatorial mounts that are very good value and give a fairly vibration free image.

One of the things that anyone buying a telescope should know is how they differ in a property known as focal ratio – usually abbreviated to f. The f number of the system determines two properties, its field of view and how responsive the system is to incoming light. The focal ratio can be worked out simply as it is merely the product of the focal length of the lens (how long the light path from the lens to the focal point where all the light rays come to a point) divided by the aperture or diameter of the lens. So if we have a focal length of 600mm and an aperture of 80mm then the focal ratio will be 600mm ÷ 80mm = 7.5 or a focal ratio of f 7.5. This will render a wide field of view. If the aperture is 150mm and the focal length is 1500mm then 1500mm ÷ 150mm = 10 so the ratio will be f10. The field of view will be smaller than the f7.5 but the larger focal ratio has the effect of 'magnifying' the image slightly and curtailing the amount of sky seen around the object. This is advantageous in some instances.

There are two main types of telescope; the refractor, which uses a lens or combination of lenses in one cell placed at the front of the telescope and the reflector, which uses a parabolic mirror to collect the light and focus it on an eyepiece. Reflectors usually have greater aperture than refractors; they are cheaper to produce and the market is well supplied with them. Another common telescope is the mirror/lens system of the Schmidt Cassegrain and its derivatives such as the Maksutov telescope.

The refractor uses a lens, placed at the front of the tube. In astronomical parlance this lens is known as an 'object glass'. Light enters the lens, travels along the tube to an eyepiece placed at the focal point which then magnifies and clarifies the image, A good quality refractor in the 100-120mm range is a very versatile instrument and will provide a good platform for observing most deep sky objects. Refractors of this size, if a good quality instrument, will provide much better images than reflectors. A good refractor is a versatile instrument but the difficulty of making quality objective lenses in sizes larger than 150mm for commercial sale has always provided the amateur with a problem of aperture.

To see really faint and indistinct objects, a reflector is the instrument of choice as they are very durable, portable despite their larger size, and the sheer aperture and light grasp is advantageous to those looking for fainter objects or more detail in the brighter ones. Reflectors therefore become the main telescope of most observers as large apertures can be purchased for a fraction of the cost of a top quality refractor. Most reflectors are built according to the Newtonian design where a parabolic mirror at the base of the telescope reflects light back up the tube to a mirror angled at 45 degrees (a flat) and then out through the side of the tube at a comfortable height for viewing. For the price of a good quality refractor you can buy a 250mm or 300mm reflector on a Dobsonian mount.

One difference between refractors and reflectors in practice is that a refractor is 'ready to go' virtually after set up, whereas a reflector may take some time cooling down to the external temperature before its obtains fine images. Tube currents play a vital role in visual astronomy and it is best to let a reflector settle before attempting to view any fainter objects on a target list. In addition, many of the Newtonian reflectors found for commercial sale are not built for photography but just visual observing. Whilst this is not a concern for most observers, this is something to be taken into account if one uses a large reflector.

Eyepieces

A good quality eyepiece can make a huge difference to your observing experience. The eyepieces usually supplied with telescopes are quite cheap and do not bring out the best in the instrument and one should change them as quickly as possible for good quality ones to bring out the best. When buying a small telescope, do not believe the claims of the manufacturer that this instrument will magnify up to 400 or 500 times. Such claims are almost fraudulent, as at such magnifications only a blur will be observed through the eyepiece. As a general rule, a telescope is performing at its optimum when it has a magnification of x25 per 25mm of aperture. Therefore, if you have a 100mm telescope, the maximum it should be able to magnify with resultant clear detail is x100. Following this advice will forestall any frustration you will ultimately experience if you have the misfortune to be sucked in by the advertiser's gimmicks.

Like any telescope, an eyepiece will come in a choice of focal lengths. These are usually displayed on the barrel as 32mm, 25mm 20mm etc. These figures

give the user the focal length of the optical elements within the eyepiece. Each eyepiece will perform slightly differently on each telescope that the observer uses and the stargazer will have to calculate the magnification obtained by each eyepiece on each telescope they use (if you are lucky to own more than one!). The magnification of any eyepiece can be obtained by remembering that the focal length of the eyepiece (in mm) divided into the focal length of the telescope (again in mm) gives you the magnification with that system. So if one has a 1000mm focal length telescope and a 32mm focal length eyepiece then 1000mm ÷ 32mm = x 32 or magnification of 32 times. The image opposite shows a typical selection.

It is worth remembering too that a small telescope is not going to show you the wonders of the universe on the same scale as a large observatory or the Hubble Space Telescope will do. Many objects show all kinds of details but most do not. Learn to work within the confines of what your equipment is capable. A small telescope should be able to see all 110 objects in the Messier catalogue and a host of brighter NGC objects and lots of binary and variable stars. They will show everything mentioned in this book.

Sun, Moon and The Planets

One would expect the Welsh to have the names of the planets somewhere in their culture. Although modern Welsh planet names follow the traditional pattern of Romano-Norse names that are tied in with the days of the week, the Welsh gods and goddesses associated with each name have different attributes to the conventional deities. The names for each astronomical object along with their current English (Norse) names are:

Planet/Object	Welsh Name	Meaning	Norse God	Welsh God
Sun	Haul (Sul)	Sunday	Sol	Llew
Moon	Llun	Monday	Mani	Ceridwen
Mars	Mawrth	Tuesday	Tiw	Nudd
Mercury	Mercher	Wednesday	Wodin	Llew
Jupiter	Iau	Thursday	Thor	Taranis
Venus	Gwener	Friday	Freya (Frigg)	Gwene
Saturn	Sadwrn	Saturday	Tyr	Tegid

As far as the Welsh connection is concerned much of the above may be interpretational as any reading of Welsh mythology will show, several deities are responsible for each astronomical object and many myths are associated with each figure. Principally though, the Sun and Moon are governed by several gods and goddesses so let us examine the links for each.

Moon

The Moon is dominated by Ceridwen, the mother of the bard Taliesin, but other Welsh deities such as Arianrhod, Don and Blodeuwedd are also involved with its waxing and waning with Arianrhod being responsible for the full moon. The Moon has always been seen as a female entity.

The familiar shape of the 'Man in the Moon' is totally different to the ancient Celts, who see it as a hare or rabbit in the moon. This is due to the connection between Ceridwen and this animal. In a charming myth, Ceridwen was disappointed by her child Morfran, who was not as good looking as she wanted him to be and was of a dark and ugly disposition. She then heard of a potion that could change her son into a beautiful creature, so she gathered the herbs, got her magical cauldron and lit a fire ready for the brew. Unfortunately, she found that the ingredients had to be stirred in the pot for a year and being the goddess of the Moon, had not time for this.

She employed two young boys to do the job for her, including a child called Dylan, but he got rapidly bored and wandered off leaving the other boy, Gwion to do the job. After stirring and mixing for days on end, Gwion was exhausted and fell asleep. Upon waking, he found the cauldron hissing as the mix was drying and in his haste to get water, his cloak caught the cauldron and pulled it off the fire, spilling the contents across the ground. As he ran after it, trying to save the contents, the last drops fell onto his hands, burned him and he stuck his hand in his mouth, swallowing the last of the potion. This did not make him beautiful as it was not ready, but it gave him the power to change into any form he wished.

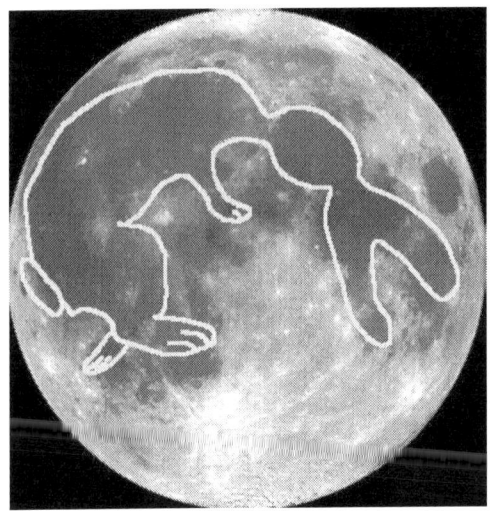

Ceridwen heard the commotion and came to investigate, but Gwion took flight across the fields; she saw him and shouted to him. Afraid, Gwion changed into a hare and ran off. Ceridwen used her magic to change into a fox and gave chase. Catching up with Gwion, she just missed him as he jumped into a river and became a salmon and swam away swiftly, she jumped in and became an otter and began to catch him. Knowing that he would be caught, Gwion leaped into a field of wheat and became an ear of wheat. Ceridwen arrived but could not see Gwion, so she changed into a greedy hen and ate all the wheat in the field, including poor Gwion.

The magic in the potion affected Ceridwen and she became pregnant; months later she gave birth to a beautiful baby boy with a "Shining Brow" or beautiful countenance, and so she called him shining brow, or Taliesin in Welsh. Ceridwen and Gwion are inextricably linked as the shape of the hare on the moon when it is full.

Sun

The Sun is chiefly accounted for by Llew, the son of Arianrhod as he dips to the lowest point of the year at midnight on the 1st May and begins to rise at midnight on the 1st August, thus his 'disappearance' brings the summertime, the ninety-two days between Beltane and Lughnasadh. The god Beli also has great oversight of the Sun as his name appears in the word Beltane though many sources now claim that he also has a part to play at Samhain, the start of the Celtic year. Beli or Belenus as he in known in other Celtic cultures was worshipped across Europe and is associated with horses, perhaps like the god Apollo in Greek myths who had a chariot of the sun drawn by horses.

Mercury

The movement of the planet Mercury in Welsh usage was the province of Llew as the planet has very close links with the Sun.

Venus

This planet is commonly associated with Branwen the sister of Bran the Giant, though how ancient this particular idea is can be loosely interpreted. More usually it is associated with the goddess of love Gwene or Gwener but this may also be an interpretation from Arthurian sources and may be a shorthand version of Gwinevere.

Mars
This planet is associated with Nudd Llaw Ereint, the mythological father of Gwyn ap Nudd who becomes the leader of the Annwn or the Welsh Otherworld. Nudd is the father of King Lud, the founder of the city of London (Lud dain in Welsh) and the area near St Paul's cathedral, Ludgate, is named after him.

Jupiter
This is the almost forgotten god Taranis from the Welsh word *taran* meaning 'thunder'. In some late Roman inscriptions found throughout southern Britain, Taranis is identified as Jupiter and has his abode in the sky and governs the weather. It is possible that this deity was brought over in the Iron Age from Gaul and accepted as part of widespread Celtic worship by the time of the Roman period.

Saturn
This planet is associated with the mythological figure Tegid and is part of the legend of Ceridwen as he was her husband, also known as Tegid Foel who becomes the father (by proxy) of the bard Taliesin. In mythology he was the god of Lake Bala in north Wales

These then are the names of ancient planets and the two brightest bodies in the sky (the Sun and Moon). It must be said that their associations with the planets may well be very loose and open to other interpretations, especially as so many of these putative gods and goddesses cross over in many Welsh tales.

The additional planets in the solar system are Uranus – Wranws; Neptune – Neifion; and Pluto – Plwton, but as these were not known to ancient sky watchers their names are more modern and almost direct translations from English.

In this book we shall not deal with the movement of the planets through the constellations as so many of the figures they represent appear in the constellation tales. It is always instructive to find where the planets are in the sky and in which constellation they are located as they do move about relatively swiftly. Any one of the monthly star charts available on the internet or the monthly run down of the sky posted by the Brecon Beacons Observatory on its Facebook page at: https://www.facebook.com/BreconBeacons Observatory/1633513230212627/?ref=aymt_homepage_panel will give the reader the required information to follow the sky on a regular basis and is a source of information and pictures of various objects taken at the observatory.

Meteor Showers and Aurora

When one is observing you can occasionally see a bright meteor or 'shooting star' burning up in the high atmosphere. Most of the bright streaks of light that one encounters are no more than the size of a grain of sand, yet these tiny particles are speeding along at tens of km per second and therefore their ablation in the atmosphere makes them burn brightly.

Most of these bodies are very small but on occasion a very bright meteor or Bolide survives long enough to make it all the way to the Earth's surface, where it can be examined. There is a rather fine distinction in the naming of such bodies. If a body is outside the atmosphere it is a meteoroid, coming through the atmosphere it's a meteor and if it is found on the ground it is a meteorite. Most of these bodies disintegrate about 30-50km above the observer's head so there is very little danger in watching a meteor shower. Many meteorites can be seen in local museums and they can be bought online from various collectors.

Most of these bodies originate in the asteroid belt or in comets. The majority of the meteors seen in fact have cometary origins; comets leave behind tiny particles of dust as they round the sun. As this dust is left behind in the orbital trail of the comet, occasionally the Earth's orbit intersects with this debris and a meteor shower results. When one is observing a shower the meteors seem to come out of a particular point of the sky known as the radiant. These areas are always in one constellation or another and so the showers are named after the groups from which they appear to originate. Meteors also have different colours depending on their origin and chemical compositions; some are distinctly green or yellow whilst others are white.

The greatest and most productive showers are associated with known comets: the Perseids have a parent comet named Swift Tuttle which has a 120 year orbit, the Geminids have a rocky body known as Pheathon 3200 with an orbit of 1.3 years whilst the Leonids are associated with the comet Tempel-Tuttle which has a 33 year period. The Orionids and the η Aquarids are part of Halley's Comet with a 74 year period. There are several such showers per year and they vary in interest and the number of meteors that can be seen by the observer.

The best showers of the year are shown here. The Zenithal Hourly Rate (ZHR) is a corrected function that attempts to take into account any fainter meteors not noticed by the observer and is therefore a little skewed. As a guide if one is going to watch a shower, then half the number contained in the ZHR represents a more likely number of meteors per hour that one can spot.

Name	Peak Dates	ZHR
Quadrantids	3rd January	40
Lyrids	22nd April	15
η Aquarids	5th May	20
Perseids	12th August	80
Orionids	21st October	30
Leonids	17th November	30
Geminids	13th December	60

Some meteors can have long trails, have terminal bursts as the material 'explodes' and others 'sputter' with mini bursts as they enter the atmosphere. Some observers have also reported hearing whooshing sounds or crackling from the entry of meteors but it depends on the observer and their acuity.

Watching a meteor shower is a great activity, though it is often punctuated by groans of dismay as some observers see a bright meteor and others looking in different directions have missed it. The trick to see as many as possible is to lie on the ground! As humans we have stereoscopic vision that stretches 180° in left-right and up-down directions so if one lies on the ground one can see the entire sky. Get a camp bed or lounger, wrap up warm in a sleeping bag and record the trails of meteors on a print off of the night sky that evening. Watch the meteor trails and record their path through the constellations and note the time of their passage. If many reports from observers in different localities are correlated then the height of the meteors can be established by trigonometry and many more details and counts can be realized. This adds to our knowledge of such events and makes for an informative and enjoyable experience.

Another activity that requires very little other than the naked eye is observing aurora. These are fairly rare events from mid northern latitudes but the appearance of an auroral curtain is a wonderful sight and is one of the easiest subjects to photograph.

The aurora is named so from the Latin for 'sunrise' as ancient peoples saw the glow as heralding the rising of the Sun. Aurora is also the Roman goddess of the dawn. The aurora is caused by particles from the solar wind impacting on the magnetosphere of the Earth. The magnetic field lines bend and some particles make it through to where the magnetic field lines concentrate them at the poles.

These particles sweep into the upper atmosphere of the Earth and produce shimmering curtains and rays of light upon their interaction with the

molecules of the air. Generally, aurora happen at altitudes of 50 km or more but can vary in depth. Although technically the aurora can be seen in many latitudes, their concentration at the magnetic poles makes them a rare event for lower latitudes and one needs to travel to Canada, Iceland or Norway see them in their full glory as they occur in auroral zones which are typically about 5° of latitude and within 20° of the magnetic poles.

The aurora can therefore be seen in two places on Earth usually – the north and south poles. The northern apparition is known as the Aurora Borealis and the southern apparition is known as the Aurora Australis. The usual colours of the aurora, green, red and blue, are indicative of the chemistry of the atmosphere as green and orange-red come from oxygen at different heights and densities (with a little from nitrogen too) whilst red and blue is primarily the signature of nitrogen at differing heights and densities.

The aurora will only move into lower latitudes if there is a particularly large storm on the sun resulting in a Coronal Mass Ejection (CME) where a huge amount of mass (1015 kg) is ejected at high speed (around 700km/s) from the Sun and then crosses space to impact directly on the magnetosphere of Earth. Such events are not particularly rare but they have to be directed at the Earth to achieve the maximum effect. Once seen, the aurora is a magical sight that is long remembered.

Deep Sky Catalogues and Types of Deep Sky Objects

The starry heavens not only contain stars, but also patches of diffuse light known as nebulae. This word is Latin for clouds, and some of these objects do indeed resemble clouds to the naked eye. Until the advent of the telescope, anything that could not be resolved with human vision was referred to as nebulae. Today we know these nebulae to be made up of several different objects, star clusters, gas clouds and galaxies. These vague objects were catalogued by several different observers, but only two catalogues have survived in general use amongst amateurs; Messier's catalogue and Dreyer's *New General Catalogue*.

Of these two catalogues, that compiled by the Frenchman Charles Messier (1730-1817) is the best known. His list of 110 objects contains the finest deep sky features that can be seen from Earth, by amateurs. Working in conjunction with his friend Pierre Mechain in the latter years of the eighteenth century, Messier compiled this catalogue as a list of 'objects to avoid', due to his passion of searching for comets. Most of the objects in the Messier catalogue could well be mistaken for comets in the early telescopes of his day, as their resolution left much to be desired. Today his catalogue is the best introduction to deep sky observing there is. It contains every type of nebulous object including planetary nebulae, galaxies, star clusters and more. During his lifetime, Messier discovered twenty-one comets, a remarkable achievement for any man, but there is no comet Messier to honour him, instead his fame lives on in his catalogue of 'objects to avoid'.

In a star atlas, and in this book too, Messier's objects are referred to by the designation 'M' followed by the number in his catalogue. Messier was not too concerned with placing them in meticulous order, they were numbered as he discovered them, thus M1, the Crab nebulae, is to be found in the constellation of Taurus, whilst M2 is a globular cluster in Aquarius, M3 is a globular cluster in Canes Venatici, and so on. Messier was going to compile the list in right ascension order eventually but ill health and his eventual death prevented him. Right to the end, Messier was still collating observations for his catalogue as so many nebulous objects were being found. The catalogue also makes no distinction as to the nature of the object either but Messier's objects will be discussed in full in the notes on each constellation.

In contrast to the seeming chaos of the Messier catalogue, the efforts of John Dreyer (1852-1926) in compiling his *New General Catalogue* is a work of art. The catalogue and its objects were observed by William and John Herschel in the eighteenth and nineteenth centuries, with John travelling to Feldhausen in South Africa to complete his father's work by compiling observations of nebulous objects across both hemispheres of the sky. Their combined catalogue became known as the *General Catalogue of Nebulae and Star Clusters* but with Dreyer's orientation of each object in right ascension order it became referred to as the *New General Catalogue*; affectionately shortened to the NGC. There are over 7000 entries in the NGC, but the great majority are relatively faint or obscure objects compared to the Messier catalogue.

Like Messier, they did not make any distinction between types of objects in their numbering system but they did utilize telescopic descriptions of the objects therein portrayed in a shorthand manner that is easy to pick up. The reader is best referred to the NGC to learn this shorthand. Nonetheless, these two catalogues are the mainstay of amateur observational astronomy today. Such star charts as *Sky Atlas 2000* use both systems of reference, as does *Norton's Star Atlas*. Most amateurs refer to deep sky objects by their Messier connotations or by a well-known name if the object was outside Messier's viewing range.

As each deep sky object can be broken down into type, perhaps it is instructive to explore these briefly.

Nebulae

Nebulae are now separated from their older appellation for everything non-stellar into proper descriptions. They are generally clouds of gas or dust that can be found in profusion along the Milky Way. They mostly consist of hydrogen gas and carbon or silicate dust, and eventually under external and internal currents and pressures, collapse to form stars. Most nebulae contain a cluster of hot young stars that are the result of such a collapse. Due to the intense ultraviolet light of these stars, the nebula shines in a process of florescence known as *ionization*, and such nebulae are sometimes called ionized hydrogen regions (HII). Nebulae come in three forms, emission nebulae, reflection nebulae and dark nebulae, the difference between these classes as far as an observer is concerned, is simply in light output. An emission nebula gives out light due to the intense ionization process, and like stars, will shine by their own light. A reflection nebula is merely a cloud of gas or dust that shines by reflecting the light of nearby stars and are generally dimmer than emission nebulae. Dark nebulae are simply gas or dust that has yet to collapse or is not illuminated in any way but rather, blocks the light of background stars.

Galaxies

Formerly, objects called 'spiral nebulae' were found to exist in profusion all over the sky but lack of resolution and an appropriate distance scale hampered astronomers from discovering their true nature. In 1926 they were found to be galaxies in their own right, lying outside the realms of the Milky Way. The famous astronomer Edwin Hubble, determined the distance of many and also found that the galaxies were also receding away from us – the universe was expanding. Astronomers catalogue galaxies in harmony with their visual appearance according to the list here and the simplicity of this grouping has survived the test of time.

- S – Spiral Galaxies
- SB – Barred Spiral Galaxies
- SO – Lenticular Galaxies
- E – Elliptical Galaxies
- Irr – Irregular Galaxies

Within the realms of our home galaxy there are likely to be only four types of object described. These are nebulae, planetary nebulae; open or galactic star clusters and globular clusters. The meaning of these phrases is given below, and hopefully described simply so that the beginner can easily grasp these terms.

Globular Clusters

Globular clusters are distant, spherical or globe-like assemblies of many thousands of stars, much richer and more compressed than the open clusters. Initially discovered by William Herschel in the eighteenth century, it was the American astronomer Harlow Shapley (1885-1972) who made a study of these objects and found that they were comprised of population II stars and are members of the galactic halo. There are approximately 150 globular clusters orbiting our Milky Way galaxy. Globular clusters do show some marked differences between each object even through a small telescope so try and mentally note these if you can.

Planetary Nebulae

Planetary nebulae are the end product of stars that have a particular range of masses. Generally, PN result from the deaths of stars that are sun-like in mass up to about 5 solar masses. As they die they slough off the outer layers of the distended red giant that they have become to leave behind a gaseous shroud with a bright point-like white dwarf shining away in the centre.

The general shape and appearance of planetary nebulae is one of the most obvious features that observers will see. Planetary nebulae have some beautiful shapes that are almost unique to each of them; the morphologies or shapes they possess are quite stunning, defining the nebulae well and resulting in some of the common names for these objects such as 'ring' nebula or 'spirograph' nebula.

Star Clusters

These are groupings of stars that are held together by their mutual gravitational attraction and have been born from a single cloud of hydrogen and helium gas in the arms of spiral galaxies. They are generally known as open or galactic clusters are very representational of some types of stars in a galaxy.

This completes our observational run-down of the type of heavenly object to be discerned through a telescope. Such items as quasars, x-ray binaries, supermassive black holes and a whole host of other objects are beyond the scope of this book as specialist (and professional) equipment and specialized techniques are required to observe them.

Watching the sky can be a magical, if not mystical experience. Go out and enjoy the Welsh tales of the constellations and follow up on some of the wonderful objects the sky has to offer no mater what optical aid you use.

CHAPTER TWO
The Autumn Constellations

Eternities before the first-born day,
Or ere the first sun fledged his wings of flame,
Calm Night, the everlasting and the same,
A brooding mother over chaos lay.
And whirling suns shall blaze and then decay,
Shall run their fiery courses and then claim
The haven of the darkness whence they came;
Back to Nirvanic peace shall grope their way.

So when my feeble sun of life burns out,
And sounded in the hour for my long Sleep
I shall, full weary of the feverish light,
Welcome the darkness without fear or doubt,
And heavy-lidded, I shall softly creep
Into the quiet bosom of the Night.

— James Weldon Johnson

Chapter Two —
The Autumn Constellations

The longer nights of autumn are tinged with the promise of frosty evenings to come, but are filled with many constellations of interest. Autumn is permeated with large amorphous constellations, the patterns of which may be a little difficult to discern with the untrained eye. The groups of Cetus, Pegasus, Andromeda, Pisces, Aries, Triangulum and Aquarius create a wonderful panorama, with the more southerly constellations of Capricorn, Sculptor, Fornax and Pisces Austrinus fading rapidly from view.

Although the autumn constellations are large, they do not contain a corresponding amount of deep sky objects. That is not to say that there are none, after all, the great galaxies of Andromeda and Triangulum, the closest large galaxies of the local group are outstanding objects only available for examination at this time of year. Rather it is the general blandness and lack of bright stars that have caused astronomers to give autumn the appellation 'the dismal season'.

Thankfully, at this time of year the summer constellations are still accessible and make a pleasing counterpoint to the indistinct stars of autumn. The glories of Cygnus, Lyra, Vulpecula and Aquila are still within reach of the hardy observer. The nights of autumn are long and cool, and the melancholy of the dying grasses and trees, loss of wildlife as they seek to hibernate, are almost reflected in the spare autumn stars.

Autumn is a transition period between the glories of summer and the grandeur of the starry winter. As with spring, we are looking out into deep space, beyond the plane of our galaxy, into the halo and beyond. Unlike spring however, there are no huge clusters of galaxies to relish, only one or two scattered groups that provide a welcome light in the celestial gloom of this dreary season. Wrap up warmly then, and enjoy some of the vistas that autumn opens to the casual observer, as some are sights that you will remember forever.

The main group of the autumn sky is the constellation of Pegasus. This is a large square with four stars delineating the corners with the rest of the autumn groups flowing from it. Directly below Pegasus can be found the main stars of Pisces and further south the body of Cetus the Whale marks the stars down to the southern horizon. To the east of Pegasus and attached to the upper left star of the square is a train of stars of the constellation of Andromeda and off the end of the third bright star in this group is the human-like figure of Perseus. Below Andromeda are two triangle shaped groups – Aries the ram and Triangulum the triangle. If one follows the neck of the horse down from the lower right star of the square one comes to the constellations of Aquarius and Capricornus.

Using the autumn constellations as a starting point, one can concentrate on learning about the stars themselves. The easiest way to engage with this is to observe them and do some personal research into the physical qualities of the most familiar stars – those that appear brightest in our sky. An excellent source for such research is *Star List 2000*, or *Burnham's Celestial Handbook* but whether or not the bright star you see is colourful, there is a more basic reason for observing it: doing so helps you appreciate it as both an example of several general classes of stars and as an individual star.

When looking at stars we should always bear in mind not only their appearance, but the physical reality which underlies that appearance. Doing so enriches the experience, making the sight more beautiful because a holistic approach makes the observation more meaningful.

Autumn has the fewest bright stars, and is thus a good place to start our observations. If you are currently in another season, you may wish to simply

Autumn constellations

read about these autumn stars. Remember, however, that when we say 'autumn stars' or 'summer stars' we really mean just those visible in the mid-evening hours for that season. That time of night is the most convenient for a majority of people. But if you are willing to go out later in the night, you can check out the stars of the next season (maybe even two seasons ahead if you get up before a winter dawn to observe). So there is no excuse for putting off actually doing (not just reading about) these activities!

Distinctive Stars

There is only one first-magnitude star in this quarter of the heavens that is best seen while Earth's northern hemisphere is experiencing autumn, but it is low on the southern horizon. Fomalhaut, spectral type A3V is the lone bright beacon in Piscis Austrinus (the Southern Fish), or indeed in a whole vast, dim section of the heavens. It is in fact the 18th brightest star in apparent luminosity in the sky. Fomalhaut beats Antares for the title of the most southerly first-magnitude star visible from British latitudes. The true brightness of Fomalhaut seems to be a little greater than that of the summer star Altair, but at 25 light years distant it is somewhat farther away from us.

Fomalhaut is distinctive in having a companion star too far from it to be bound to it by gravity, yet lying at the same distance from us and moving through space with a motion similar to Fomalhaut's. One possibility is that Fomalhaut and the star are the final survivors of what used to be a star cluster or star association, the other members having long since drifted off on their own. The companion is nearly one light-year away from Fomalhaut and can be found south of Fomalhaut in the sky. It is an orangish star with an apparent magnitude of 6.5, so that its true brightness is only about 10 percent that of the Sun and therefore, binoculars are required to view it. Note the colour contrast with Fomalhout. Additionally, a ring of dust surrounds Fomalhaut; in 2008 an extrasolar planet was discovered in orbit around this star that has been seen directly with the Hubble space telescope. Known as Fomalhaut b, this planet has three times the mass of Jupiter.

Other stars worth noting are the binary star Gamma Andromedae, spectral type K2II, which is a golden yellow colour whilst its companion is a pale blue hue, though the noted variable star observer J.R. Hind thought of the colour of the companion as 'lilac' in comparison. In contrast to Beta Pegasi (spectral type M0III), which is orange – try to spot the difference.

Andromeda
(Blodeuwedd and Rhiannon)

In Welsh mythology this maiden represents two tales. The first is that of the woman Blodeuwedd, fashioned from the flowers of the fields by Gwydion and Math as a bride for the young man Llew (the constellation of Perseus). Although happy at first, Blodeuwedd fell in love with one of Llew's friends and confidents, Goronwy. She nagged Llew until he told her the secret of how he could be killed. The two then conspired to kill him. Once Gwydion had restored the soul of Llew and created the Milky Way in the process, he took revenge on Goronwy; Blodeuwedd was banished to the sky to rule over creatures of the night, her beauty fading like the flowers after sunset. In some versions of the tale Blodeuwedd is turned into an owl, another creature of the night.

The second tale is that told of Pwll and Rhiannon in the *Mabinogion*. This lady of the otherworld seeks a marriage to Pwll, the Lord of Dyfed and is riding a beautiful white horse that none of Pwll's retainers can catch as the beast always magically pulls ahead of them with Rhiannon riding serenely on its back. Finally, Pwll gives chase, catches up with Rhiannon and they fall in love and marry, setting out on many adventures. The horse (Pegasus) and the rider (Andromeda) make up the mythical Rhiannon and the white magical horse.

* * *

As a constellation, Andromeda contains some of the greatest deep sky treasures in the entire heavens, one of which is M31, the Great Nebulae, now known as the Andromeda galaxy. Although this galaxy lies 2.9 million light years away, this is relatively close on an astronomical scale, and the intense light of untold billions of suns enables the galaxy to be perceived by the naked eye as a fourth magnitude smudge of light at the top of a line of stars northward of δ Andromedae.

M31 was not discovered by Messier; it has been known since very early times, and many legends have grown up around this astounding object. The first person to view M31 through a telescope was Simon Marius in 1624, who described the spectacle of its soft, glowing light as if he were looking at "a candle shining through horn". In a pair of binoculars, the view is stunning, the bright milky nucleus does not show the same condensation as a globular cluster would, but is rather cloudy, and a little less opaque. The spiral arms of the galaxy can be seen as a sliver of faint luminescence radiating out in symmetrical

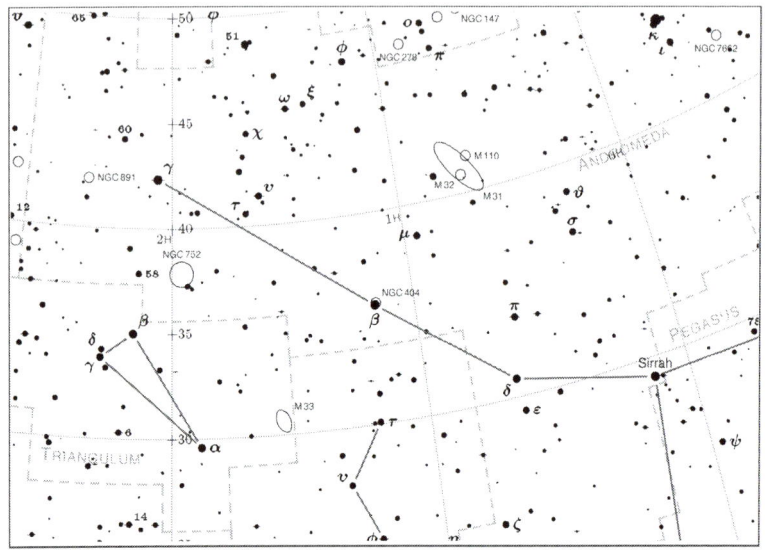

Andromeda

projection on both sides of the nucleus. Seen through a rich field telescope, the galaxy is transformed into a glowing elongated mass of soft white light that can be traced for almost one degree against the darkness of the sky.

The brightness of M31 recommended itself to the astronomer Edwin Hubble, who was accumulating evidence that the spiral nebulae were in fact different systems outside the Milky Way, island universes in their own right. During the late 1920s, Hubble began to resolve this galaxy into stars with the aid of the newly commissioned 100 inch Hooker telescope on Mount Wilson. The photographs he obtained showed several Cepheid variables that were closely examined over a period of months. The Cepheid period luminosity relationship was well established at this time, so Hubble was able to calculate the distance to this nebula. Although his conclusion of 750,000 light years is now considered in error, it was sufficient to prove that the 'spiral nebulae' were indeed galactic systems at tremendous distances from us. Hubble's findings opened a whole new universe to mankind, one that has amazed, perplexed and intrigued all manner of persons since.

Messier 31 is accompanied by two elliptical galaxies that lie very close by. The brightest of these is the companion which Messier catalogued as M32, an E4 type galaxy shining at magnitude eight, having a slightly mottled aspect as viewed in a telescope. It can be seen with binoculars as a hazy patch to the south

Messier 31

of M31, slightly fainter than the nucleus of its parent galaxy, but the brightest of the four known satellites of M31. (The other two lie over the border in Cassiopeia). The second companion can be found to the north of the mass of the Andromeda nebulae and rejoices in the name of NGC 205. It is a minor mystery how Messier came to miss this companion, as through a telescope it is in the same low power field as the others of this group. Messier certainly recorded fainter objects in his catalogue, but perhaps he was so enthralled with the most obvious object in this area, that he looked no further after noting them.

Another explanation focuses on Messier's magnification, which he universally used as x137, rendering the field small enough to miss NGC 205 altogether. NGC 205 is an E5 type galaxy of tenth magnitude that may be seen with good binoculars, and is a hazy undefined object through a small telescope.

Other objects of note within Andromeda include a attractive loose star cluster to the south of the main body of stars at coordinates RA 01h 57m 48s Dec 37°41m. This is the galactic cluster NGC 752, a collection of 125 stars in an area of sky larger than the full moon. NGC 752 can be readily seen in a pair of good binoculars, whilst a telescope with a low power ocular resolves the field beautifully. The cluster lies about 1300 light years away and shows an abundance of relatively metal-poor stars, thus making this group a rather ancient cluster.

NGC 752

In the opposite direction, to the northwest of M31, lies one of the nicest planetary nebulae in the autumn sky, one that is best seen with a telescope, but can be viewed with a pair of binoculars, although through such instruments it is an unremarkable object. This is NGC 7662, a round, greenish-blue planetary close to the star 13 Andromedae at RA 21h 01m 30s Dec16°11m. It is an arresting sight in a telescope as it shines brightly at magnitude eight, and shows a perceptible disc that is the trademark of objects of this type. NGC 7662 is fairly remote, about 5600 light years, giving the nebulae a diameter of almost one light year. The colour is a pale blue – so much so that it has been dubbed 'the blue snowball'.

Four degrees southwest of M31 is one of the easiest long period variable stars to follow with binoculars or a small telescope. It can be found in the same field as Theta, Rho and Sigma Andromedae, and is a red giant known as R. Andromedae. It has a warm orange colour that deepens as the star fades. It can be found with little effort in a pair of binoculars as a seventh magnitude object when at maximum, but it dips to less than 13th magnitude at minima in a period of 409 days. R. Andromedae is a halo object of spectral type S, which makes it a fairly rare specimen apparently lying over 1000 light years distant.

The first multiple planetary system ever found around a normal star consists of three planets in orbit around υ Andromedae. The innermost of the three planets, υ Andromedae b, is at least three-quarters of the mass of Jupiter and orbits only 0.06 Astronomical Units or AU from the star. An AU is the distance between the Earth and Sun and was the first distance scale discovered by astronomers. One AU is equivalent to 150 million kilometers. Andromedae b traverses a circular orbit every 4.61 days. The middle planet contains at least twice the mass of Jupiter and it takes 242 days to orbit the star once, residing approximately 0.83 AU from the star, similar to the orbital distance of Venus. The outermost planet has a mass of at least four Jupiters and completes one orbit every 3.5 to 4 years, placing it 2.5 AU from the star. This is a fascinating system for contemplation.

γ Andromedae is a wonderful double star for observers with a small telescope, the primary is a lovely yellow star of the second magnitude, and the companion is a fifth magnitude blaze of blue light, that is almost green in contrast with its near neighbour. Both stars are approximately 260 light years away. Observers with telescopes or giant binoculars may be able to see the beautiful edge on galaxy NGC 891, RA 02h 22m 36s Dec 42°21m lying about 5 degrees east of γ, shining at tenth magnitude. This galaxy is distinguished by its remarkable lane of obscuring dust and lies 20 million light years away.

Just to the north of the star β Andromedae, or Mirach to give it the correct name, is a lovely little elliptical galaxy NGC 404 which shines out in a modest telescope at 11th magnitude if one can remove b from the field. NGC 404 is relatively close by on a cosmic scale as its distance is 10 million light years. The white of the galaxy makes a pleasing contrast to the orange hue of β Andromedae.

Aquarius
(The Murdered Merchant)

Aquarius is a rather loose autumn constellation with very little impression of outline, yet containing a little Y-shaped asterism called the 'Water Jar' which is generally used as the constellation's centre point, from which the observer can find the rest of this expanse of stars. The name Aquarius means 'The Water Bearer' although the figure that the constellation supposedly represents has long since been lost. It is however a group of antiquity, probably being formed by the Babylonians, and taken up by different cultures since. Indeed, the Romans knew the group as the Peacock of the goddess Juno, the queen of the heavens.

A possible connection to this constellation, with its image of a man pouring water from the sky is one that reminds one of the legend of Llangorse Lake, the largest natural lake in south Wales.

The story tells us that in olden days, all the land beneath the lake belonged to a cruel princess, possibly a reference to the dwellers on the Crannog in the lake which would have been owned by a person of importance. A local farm boy saw her bathing in the lake one day and was smitten by her. Though he was very poor, the princess agreed to marry him on condition that he brought her lots of money or livestock, she suspecting that this would drive the poor boy away as it was obvious that he had nothing. However, so turned in his love for the princess and anxious to please her, the farm boy waylaid and murdered a wealthy local entrepreneur (merchant in some tales) and after robbing the corpse of its money bags he gave the gold coins to the princess.

They were meant to live happily ever after even though the farm boy felt remorse for his crime. The princess took it as her right to have the money and they married in a large ceremony. Nevertheless, the crime was not forgotten by the victim. On their wedding night, they were visited by the ghost of the merchant who wanted revenge. He warned the couple that their crime would be avenged by him eventually and a terrible disaster would overtake the ninth generation of their family. The newlyweds thought they could accept this punishment as it would not apply to them and over the years they grew to have a large family who married into the local farming community and the land they inhabited grew into a small town.

However, vengeance was waiting for them as upon the birth of the first child of the ninth generation of the couple, a terrible flood burst from the hills and sky, filling the town with water as if poured from a heavenly jug by the vengeful victim, the water drowning the land and the inhabitants, including the old couple who had been preserved from death in old age to see this vengeance. Today there are still stories in the area of the town under the Lake. Some say that the couple's children are the eels that inhabit the lake.

* * *

Aquarius is fairly rich in deep sky objects worthy of the amateur's attention plus some very interesting variable stars for those equipped with larger instruments. As it also straddles the ecliptic, occasionally some of the planets can be found adorning the field, bringing a little sparkle to this amalgam of second and third magnitude stars.

The first object on our deep sky tour is the wonderful globular cluster M2, a bright sixth magnitude array that can be seen in binoculars some 6 degrees north of Beta Aquari. M2 was discovered by Messier in 1760, soon after he began to formulate his catalogue, although it seems that Giovanni Maraldi was

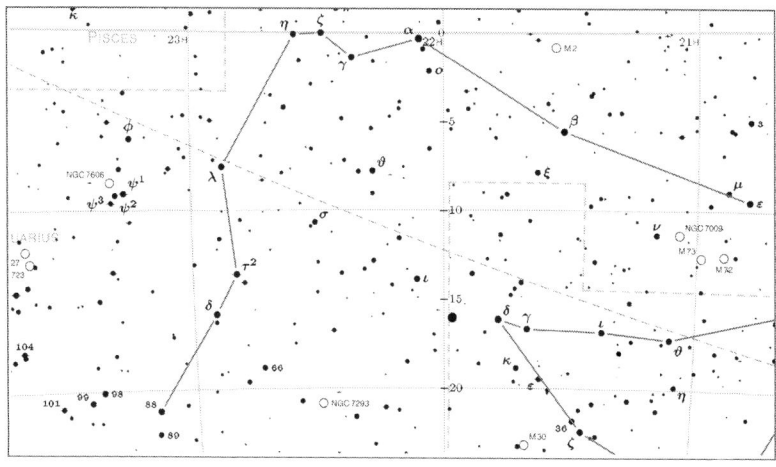

Aquarius

the first person to see it some 16 years previously. M2 is a fine object in a small telescope, it is a rich round blaze of white stars, which shows a little resolution along the edges, but the centre is an opaque mass of some 100,000 stars shining at magnitude 14 or 15. M2 lies over 50,000 light years away, and shows the lowest proper motion of any globular in the sky, possibly due to it reaching the furthest point in its orbit away from the galactic nucleus, and is now preparing to fall back into the galactic disc, although evidence of this scenario is scarce.

Another globular catalogued by Messier is the faint cluster M72, a tenth magnitude rather bland object in the south of the constellation. It is barely visible in binoculars, but no detail is really revealed through a small telescope either, it simply remains a round haze of white light in a barren field.

M72 apparently lies 54,000 light years away. Aquarius also holds another Messier object that is something of a curiosity. This is the object called M73, discovered a few nights after M72. It is not a cluster or nebulae, just a faint group of three unremarkable stars in an otherwise sparse field. Messier noted them as such, but added that they lay in a field of nebulous gas, though this must have been a shortcoming of his telescope, as no nebulae lies at this position.

One of the showpieces of the autumn sky can be found in the same general area of M72 and M73, though how Messier came to miss it is not explained. This object is the planetary nebulae NGC 7009, a beautiful object bursting with detail in the field of a small telescope, and bright enough to be seen in binoculars as a fuzzy star.

Messier 2

NGC 7009 was called the Saturn Nebulae by Lord Rosse (1800-67), the famous Irish observer who also gave M1 and M51 their familiar names of the Crab and the Whirlpool. A glance at this object is enough to tell you why it he called it so. The nebulae is elongated and appears to have a ring of material surrounding the nebulous disc, exactly like the famous ringed planet. The nebulae has a high surface brightness that will take high powers well, it shines at magnitude 8 and is some 3900 light years away. Observers with large instruments may even be able to distinguish the bright rays or *ansae* that flash out from the central nebulae and across the ring.

Another planetary nebulae lies well to the east of NGC 7009 and is an equally famous object of its type. This is the so-called Helix Nebulae, or NGC 7293 as Dreyer recorded it. NGC 7293 at RA: 22h 29m 36s Dec -20°48m, has the distinction of being the largest planetary nebulae visible to us. It is slightly smaller than the full moon and has an integrated magnitude of six. This should appear as a spectacular object, but its light is spread over such a large area that it rapidly diminishes, and the nebulae is practically invisible as a result. It is one of the few objects where binocular observers have an advantage, as the wide field of such instruments will gather all the available light to build up a picture of a ghostly blue-white ring of gas, although no detail will probably be seen.

If you are the owner of a wide field telescope, or giant binoculars, then give NGC 7293 a try, but for many small telescope users, regretfully the advice is to look elsewhere. The Helix is large merely because it is the closest planetary nebulae to our solar system, lying at a distance of 85 light years, giving the shell of ejected gas a diameter of 0.3 light years. The science fiction author Robert Silverberg speculated on the possibility of rapidly mutating, radiation eating life forms in such a fierce system as NGC 7293 in his book *Tower of Glass*.

One of the strangest long period variable stars lies in Aquarius, and is known as R. Aquarii, a pulsating red giant star of class M, lying close to the naked eye star omega Aquarii. It is unusual in that on some occasions the regular pulsations stop and the star remains at minimum light. R. Aquarii usually varies between magnitude six and magnitude nine in a period of 386 days; it should be visible to binocular observers at all times. As complete reports and light curves for R Aquarii are not readily available, why not try plotting your own, and sharing the results with others of similar interest. Such work is most important, and could even form a thesis for university study.

Aquarius also contains over a dozen galaxies that unfortunately do not reveal themselves to common instruments as their magnitude averages 12.5. However, the delightful objects described above will draw the attention and affection of the observer time and time again to this lovely group of the autumn sky.

Aries
(The Welsh Sheep)

The constellation of Aries is a collection of three fairly bright stars in an area of sky to the south and east of Andromeda. These three stars are surrounded by a large number of fainter constellation members within a border that seems far too big for the little constellation it contains. Aries was introduced by the Babylonians as the figure of a celestial ram. Other civilizations have imitated this and the Greeks thought of the group as the ram that carried Phryxis and Helle across the sky away from the angry goddess Ino. Unfortunately, Helle did not hold on and fell into the sea and drowned, though the sea is now named after her – the Hellespont.

Where would Wales be without its sheep? They are a ubiquitous part of the landscape in Brecon and farming of sheep is a tradition that dates back centuries in this area. The sales of livestock in Brecon Market and the trading of wool from the fleece has made many farmers financially comfortable, the fleece being almost literally turned into gold once it was sold at market, so the glittering golden fleece of the celestial ram is a reminder of the agricultural worth of these common animals.

Sheep are mentioned a number times in Welsh tales; in Culhwch and Olwen the seven knights of King Arthur are stopped in their search for the house of the giant Ysbaddaden by a flock of sheep that stretched further than the eye could see. The sheep belonged to Ysbaddaden but were tended by the shepherd Custennin, the disenfranchised brother of Ysbaddaden who went on to help the knights in their quest to win Olwen from her father.

* * *

Aries contains very few objects of interest to the casual observer. There are few galaxies, but even for observers with large telescopes not many deep sky delights to enthrall the astronomer in this barren and dismal constellation. Aries does contain some stars that are good targets for those interested in binary systems, but on the whole Aries is best skipped over if you are looking for something to really get your teeth into.

γ Arietis is one of the nicest coloured doubles available to the amateur observer; sadly, it cannot be seen through a pair of binoculars. It consists of two magnitude 4.5 stars that are well separated. The primary is a hot blue star of spectral type B, and its slightly fainter companion is an A type star that some observers have described as yellow, lilac or white. It is a famous star in that it was the first double star ever to be discovered, being found by Robert Hooke in 1664. Hooke was not a brilliant astronomer, but he was a good

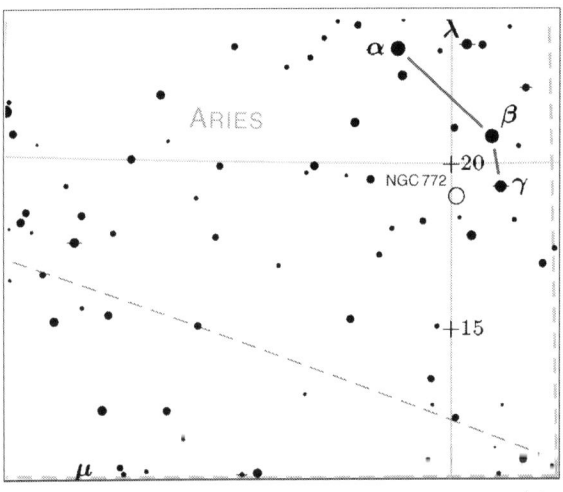

Aries

optics maker, and invented the microscope. He found the star whilst observing a comet in 1664 and brought this celestial wonder to the attention of the recently formed Royal Society.

Another lovely coloured double star is 30 Arietis, a rather difficult object to detect visually in our light polluted times as the primary object shines at magnitude 6.2, and lies in a rather stark area of the sky. Both stars are dwarfs of spectral type F, but for some reason, the companion looks almost blue in colour compared to the yellow-white primary, probably some tricky optical effect causing this colour distinction.

Aries contains another faint star of interest to observers, which can be examined by those equipped solely with binoculars. This is 53 Arietis, a star with a very high proper motion that has been dubbed the 'runaway star'. Examination of its motion has led astronomers to conclude that 53 Arietis has been ejected from the area of the Orion nebulae, M42, possibly as a result of a catastrophic supernova explosion. Two other high velocity stars can be traced back to this region, AE Auriga and Mu Columbae, causing each of these three stars seemingly to fly off along three points of the compass: west, north and south. What may possibly account for their motion is a matter of conjecture. All three stars are B type objects, obviously fairly young stars. If they were members of a system where the massive primary exploded, then the resultant mass loss by the detonating star could have resulted in the orbital 'bonds' becoming broken, and the stars then being hurled away in different directions.

Aries contains nothing further of interest to the modestly equipped amateur observer.

Capricornus
(The Sea Cow or Mermaid)

This is a large but well-defined constellation to the east of Sagittarius, and is the next zodiacal group in which planets may be well seen. Capricornus is something of a mystery; its name has always been given as the 'Sea Goat', and on old star maps it is depicted as a strange mixture of Goat and Fish. It is an ancient constellation, again being formed by the Babylonians, and being identified by the Phoenicians as their god Dagon, who was half man half fish. The Greeks applied this group of stars to their god Pan, another half and half creature, in this case a combination of man and goat; it was in this form that the librarian Eratosthenes recorded it at the Alexandrian library in the second century BCE. Chinese astrologers knew it as the 'Southern Gate of the Sun',

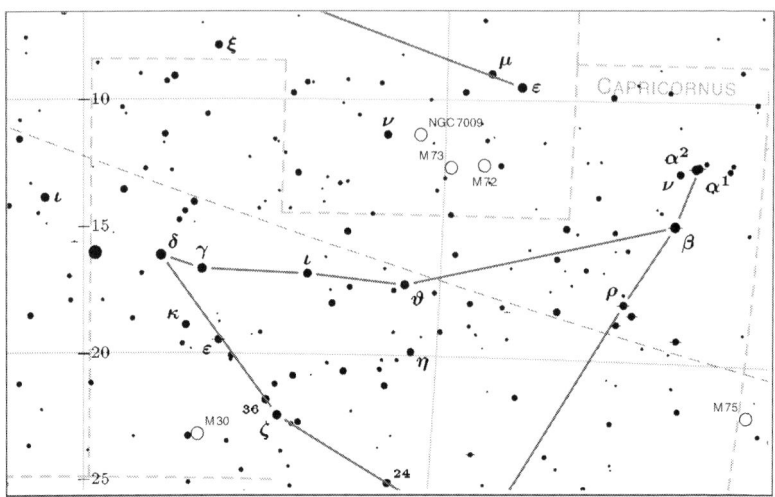

Capricorn

a reference to the fact that the constellation contains the point where, at the Winter solstice, the Sun begins its move northward.

To the inhabitants of ancient Wales this constellation represented the Morvugh or sea cow, possibly a representation of the grey seals that are common along Welsh coasts and have been associated with the creature the Scots know as the Selkie, a seal that changes shape to become a human as the word is an old name for the common seal. Perhaps tales of mermaids and mermen have come from this constellation due to the seal's ability to emulate a man or woman. Another aspect to the Welsh mermaid mythology is that of the fairy washer-woman. This form of water fairy is to be found in legend near to streams and rivers, performing her ablutions – sometimes viewed as a vision of death to come, particularly if washing a shroud or armour.

The constellation with its head and body and fish tail can also be associated with the Welsh tale of the Cwn Annwn – the dogs of the Otherworld owned by Gwyn ap Nudd. One of them, the dog Dromarch, has a single head, two front legs, and a body that narrows rapidly from the chest and terminates in three fish-like tails, very reminiscent of the common sky creature we take as Capricornus. This dog, along with its brethren hunt the cwms and lakes around the peak of Cadair Idris. Indeed, Dromarch's habitat is that of the clouds and mist according to The Tale of Gwyn ap Nudd and Gwyddno Garanhir, from the *Black Book of Carmarthen* and the dog probably has a changeling bent, being at home on land, lake or cloud.

The Autumn Constellations

* * *

Capricornus looks like the letter 'V' lying on its side with the open end pointed towards Sagittarius. It is a rather disappointing group as far as the deep sky observer is concerned, as it only contains one object within easy reach of average instruments. The constellation does however contain some interesting binary systems that are easy objects for either binoculars or a small telescope, plus some long period variables that lie within range of a small telescope.

The first of these double stars is an easy object even with the naked eye. It is α Capricorni or Al Geidi as the Arab astronomers called it, a name meaning simply The Goat. It is a lovely sight in binoculars as the two components are well separated and are both yellow stars, although the colours may not be apparent due to their low elevation. The pair does not form an actual binary system; the connection is a line of sight effect, as the component α1 is over 500 light years away, whereas α2 is a relatively close 100 light years distant.

Another binary of interest is the star β Capricorni, a wonderful coloured double that may be separated with a pair of binoculars under good seeing conditions if they are properly mounted. The stars are easily visible in a small telescope. The primary star is a golden yellow colour shining at third magnitude, whilst the companion is a lovely, dazzling white sixth magnitude object. The pair lie about 150 light years away.

Messier 30

The only deep sky object of curiosity in Capricornus is the bright globular cluster designated M30, a bundle of light about 6 degrees south of γ Capricorni, close to the sixth magnitude star 41 Capricorni. M30 can be seen in a pair of binoculars as an eighth magnitude smudge of light in a barren field of stars, and is irresolvable in a small telescope. It is not fantastically distant as globular clusters go, being 40,000 light years away. Even the world's largest telescopes have a little difficulty in resolving the central regions, where counts of the star images show M30 to contain about 50,000 stars

Cetus
(The Changeling)

Cetus is the largest constellation in terms of area in the autumn sky, and is an amorphous collection of faint stars that mark the boundaries of the fabled sea monster that was sent to attack the beautiful Andromeda to compensate for the boasting of her mother, queen Cassiopeia. Thankfully, the hero Perseus was on hand just in time to save the fair maiden. He killed the sea monster by showing it the decapitated head of the Gorgon Medusa, thus turning Cetus into stone. Poseidon, incensed that his monster was dead, placed it in the sky in a position where it could still threaten Andromeda, and roar its disapproval at Perseus. On old star maps, Cetus is always portrayed as a whale, with huge teeth, a dog's head and generally frightful appearance, which belies the nature of these gentle creatures. Big was obviously not always beautiful to the ancients.

In Celtic tales this constellation represents a swineherd who was turned into a whale. There are very few Welsh associations with this constellation apart from repetitions of the mer-cow and mermaid tales which also feature in the constellation of Capricorn. In Welsh myths this creature is said to be the Pwca, and this association is shared by Irish and Pictish tales across the Celtic world. The Pwca is supposed to be a bringer of both good or bad fortune and they could either help or hinder rural and marine communities. The creatures were said to be shape shifters that could take the appearance of black horses, goats and rabbits. They may also take a human form, but always gave away their true unnatural form by exhibiting various animal features, such as ears or a tail.

* * *

Cetus contains a few objects of interest to the casual observer, but unfortunately, its low altitude as seen from Britain tends to dilute the brilliance of

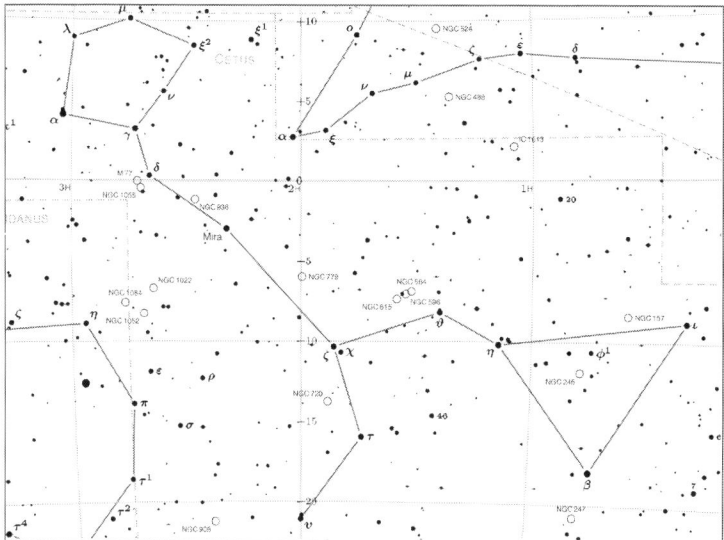

Cetus

some of them and adds one or two magnitudes to others. Identifying the group is not difficult; simply look for the head of the monster, which is the most easterly part of the constellation. Its five stars mark out a circular outline from which it is relatively easy to figure out the rest of the constellation as it spreads south and westwards. Cetus contains the beautiful variable star O Ceti, or Mira, the typical object of this form of celestial wonder, in addition to several galaxies that lie within the range of amateur telescopes.

The best deep sky object in Cetus is the Sb type spiral galaxy M77, a tenth magnitude smudge of light just under the 'chin' of the monster. It is not an easy object in binoculars, but it may be seen on a good night as a faint glowing mass of grey light 60 million light years away. M77 is a very unusual galaxy, one of the closest of a type known as 'Seyfert galaxies', after the astronomer Carl Seyfert who made a study of their ultraviolet excess and their violent nuclei in the 1940s.

Seyfert galaxies are mostly spiral types characterised by very bright nuclei in proportion to their spiral arms, and also the peculiar presence of emission lines in their spectra. Further study of these galaxies has revealed that there is a tremendous amount of energy flowing out of the core of these objects, originating from a very small space at the centre. They are also radio galaxies,

Messier 77

and some are also visible in both x-rays and ultraviolet light, evidence of intense activity the source of which is postulated to be a black hole. Astronomers think that a black hole of several million solar masses is shredding stars and gas within these galactic nuclei, and ejecting some of it into space where it collides with the intergalactic medium, creating a shock wave that causes intense radiation. Seyfert galaxies are thus related to radio galaxies and quasars, being a little lower down the energy scale.

Worth seeking out are NGC 157 and NGC 908, two galaxies with a magnitude of 11 so don't expect to see them that well as in a small telescope they will merely be little smudges of light, and all but invisible in binoculars. NGC 157 is an Sc type spiral lying 65 million light years away, which looks a little elongated in a low power eyepiece. NGC 908 is an Sc type spiral at a similar distance to NGC 157 and is a little fainter than it. Both galaxies can be viewed but their arms will be a dull haze with a faint core.

The flagship of the constellation is of course the beautiful red giant star o Ceti, or Mira as it is commonly known. Hevelius named the star and it was the only variable star known for quite some period of time. The name means 'Wonderful', and many observers will agree that it deserves the title. Mira can be seen on any autumn night even when at minimum as it varies between magnitude 4 and magnitude 9 in a period of 331 days.

On occasion, Mira becomes a lot brighter; during the late eighties and again in the noughties the star was a brilliant naked eye object shining at second magnitude, and transformed the autumn sky with its incredible orange glow that was plain to see. Its spectral type is M, and its distance is roughly 300 light years, which is relatively close for such a star. Over 4000 Mira type long period variables are known, most of which have periods between 250 and 400 days, thus making convenient distance indicators, as most of these giant stars have a similar intrinsic luminosity.

Mira is a very large star, probably around 300 times the diameter of our Sun, and one of only three stars in which spectral bands of water vapour have been found. At minimum light (a term astronomers call minima) the star switches most of its energy output into the infra-red part of the spectrum as it becomes an intense red colour and the surface temperature drops to only 1800 degrees Kelvin. Its oscillations can be followed in binoculars or a small telescope and is an ideal object to introduce the amateur to the vagaries of variable star observing. It is also a binary system with a red giant companion. The star is moving through space and leaving a trail of gas behind it as some mass is lost from the system, something that is typical of such large red giants.

A star of interest within Cetus is the third magnitude τ Ceti. It is not a binary system or variable, but is a G type star of almost the same dimension and luminosity as our Sun. τ Ceti is only eleven light years away, and due to its Sun-like qualities was picked as a target for the SETI programme, the search for extraterrestrial life. In 2012 it was postulated that τ Ceti has a planetary system, with some researchers describing a 5-planet system with one planet in the habitable zone – and area where liquid water could exist on the planetary surface. No telescope yet built will show these planets however, so we will have to await any reply to our radio signals to confirm their presence. As yet no one has answered.

One lovely planetary nebula worth noting is NGC 246, the Skull Nebula at RA 00h 47m 18s Dec -11.52m.18s. It is a large nebula that is almost 4 minutes in diameter and shines at 10th magnitude and looks a little like Pacman from the computer game. Cetus contains little else of interest to the observer with modest equipment, although owners of large telescopes will have a red letter day with the dozens of galaxies visible in this area, most of which are around 12th magnitude and are good candidates for the scrutiny of the supernova patrol. Browsing through a good star atlas will give their positions against the star of this large constellation.

Pegasus
(The Ceffyl-Dwr and the Mount of Rhiannon)

To the Welsh this was the horse of Llyr, the sea god, a beautiful white charger that figured in many Celtic myths and still has marine associations – we call the whitecaps of waves white horses. In the ancient Celtic world this constellation (and Andromeda on its back) represented the goddess Epona riding her white horse and is symbolized in the British landscape by the chalk figure of the white horse at Uffington.

We have seen from the Welsh folk tales in *The Mabinogion,* that Rhiannon, the bride of Pwll, Lord of Dyfed rides a magical white horse that has often been taken to be the white horse of the sky. The constellation is also associated with the god Belenus or Beli Mawr as he was the sun god who had horses draw his chariot. In ancient times, the sun rose at the spring equinox in the constellation of Taurus and the Horse would have rose before it, almost looking as if the horse was pulling the Sun. In this manner, the connection

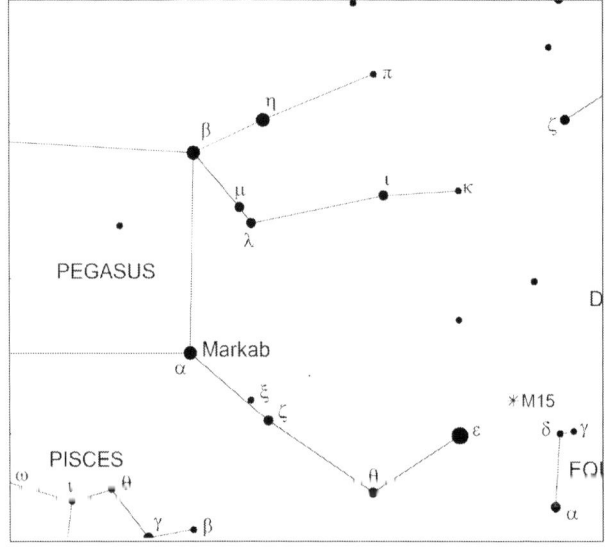

Pegasus

between the sun god, the sun and his solar, horse drawn chariot was obvious.

There is also a Brecon legend of the grey 'Water Horse' of the Honddu river, a magical horse or kelpie, what the Welsh call a ceffyl-dwr. A long time ago a weary man was lured by a small grey water-horse in the Honddu to ride him. But the playful ceffyl-dwr took him all the way to Carmarthen – far from the poor man's destination and he set off even more tired than ever! Horses are playful animals and even the Ceffyl-Dwr in the sky is upside down, perhaps to unseat any unwary riders.

* * *

Pegasus is made up of four principal stars that create the well known Square of Pegasus, the most dominant asterism of stars in the autumn sky. The neck and head of the horse are well marked with stars, ending in ε Pegasi, a lovely red star called Enif, meaning the Nose. Pegasus was placed upside down in the sky, and to complete its discomfiture, its legs have been cut off to form the constellation of Andromeda. Pegasus contains a few deep sky objects of note that can be seen in binoculars or a small telescope.

The first such object is the exquisite globular star cluster M15, which can be found by taking a line from the 'top' of the head northwards through Enif, and extending it by 4 degrees. It lies in a pleasing field of eighth magnitude

Messier 15

stars and sparkles at magnitude 7. In binoculars it resembles a roundish cometary nucleus, but the view through a small telescope is wonderful, showing partial resolution into stars, especially around the edges of this radiant ball of light. M15 is one of the best globular clusters in the northern sky, as it probably contains over 250,000 suns, lying at a distance of 42,000 light years.

In the northwest of the constellation can be found an excellent galaxy for small telescopes or even a good pair of binoculars. This is NGC 7331, at RA 22h 37m 06s Dec 34°25m; an Sb type spiral that has often been compared with the great Andromeda galaxy, M31, in form. It is an object of tenth magnitude with a relatively high surface brightness and can be viewed as an elongated blob of blue-white light. NGC 7331 lies over 40 million light years away, so may be quite a massive galaxy when compared to the luminosity of other spirals at similar distances. Several supernova have been recorded in this galaxy, so it is worth checking regularly, as such objects often outshine the galaxy from which they originate.

Pegasus contains one rather curious asterism, NGC 7772, a collection of seven faint stars that is not a true cluster, merely a line of sight apparition, lying within the southern bounds of the square. Observers with large telescopes should attempt to find a collection of galaxies close by NGC 7331, known as Stephan's Quintet, as well as more than a dozen others in this rich field of extragalactic objects.

One of the first extrasolar planetary systems was found in Pegasus in 1995.

Stephan's Quintet

Michele Mayor and Didier Queloz discovered line shifts in the spectrum of the star 51 Pegasi indicated a small, unseen companion. The planet is considered to be about half the mass of Jupiter and lies 0.5 AU away from its parent star. Since 1995, further investigation strongly suggests another planet in the system. It is also a model for the Hot Jupiter hypothesis as the planet may have formed beyond 5 AU of the star and through friction with an equatorial disc extending from 51 Pegasi, may have wandered in to its present position. Note the golden colour of this star and ponder the significance of this beautiful system.

Not too far away from 51 Pegasi is another extrasolar planetary system orbiting the star HD209458 at RA 22h 03m 10.7s and DEC +18 53m 04s. The star is spectral type G0V star very similar to the Sun, shining at magnitude 7.5. The planet is half the mass of Jupiter and orbits in a little over 3 days at a distance of 0.04 AU from the star. One of the most interesting facets of this system is that several transits of the planet have been calculated and observed via photometry with large telescopes. Those who witnessed the transit of Venus in 2004 and 2012 can well imagine the scenario if we extrapolate to this system, with its giant planet carving a huge round disc against the backdrop of the yellow parent star.

Perseus
(Llew Llaw Gyffes)

In Wales, Perseus represents the hero Llew Llaw Gyffes who was the child of Arianrhod (herself represented by the constellation Coronae Borealis) who could not prove her virginity to her uncle Math and thus had the job of massaging his feet and influencing him. She laid three curses on the poor boy, telling him that he would never have a name, never bear arms or ever be married. His uncle Gwydion thought this unfair and so he undertook the task of alleviating the curses.

When Llew was in his teens, Gwydion disguised him as a lowly shoemaker and brought him to her court at Caer Arianrhod, a place off the coast near modern day Conwy, where his skill with the bow got him named as the 'sure handed fair one' (Llew Llaw Gyffes). A few years later Gwydion again entered Caer Arianrhod in disguise with Llew in tow as a bard. During the night, Gwydion sends Arianrhod a dream in which he conjures up an army invading her court and she wakes Llew and gives him arms so that he can assist in repelling the foes. Tricked and angry, Arianrhod swears that he will never be married and so Math and Gwydion create a woman for him from the flowers of the fields, Blodeuwedd. They marry and should live happily ever after.

Many years later, whilst Llew is making laws for some of the cantrefs he holds, Goronw, one of his close companions, falls in love with Blodeuwedd. She nags him to tell her how Llew can be killed as she wants him to avoid such a fate but secretly plots to kill him. He gives her the secret, that he can only be killed if his feet are in running water. Upon hearing his secret, Blodeuwedd and Goronw plot his death. Out hunting together one day shortly afterward, they murder Llew whilst he crosses a stream, but at the moment of his death, his soul turns into an eagle – Gwalchmai or the Hawk of May – and flies into the heavens.

His uncle Gwydion knowing Llew has ascended into the sky, takes the embers of a fire and climbs into the heavens, scattering the embers behind him so that he can find the road back to Earth. The embers become Sarn Gwydion, or the Milky Way. Gwydion eventually finds Llew as the constellation Aquila and restores him to life. Returning to the mortal world, Llew takes vengeance on Goronw, killing him as he hid behind a large stone with a huge spear throw that pierced the stone. This stone, with a hole right through it, can still be seen today at Llech Ronw in Blaenau Ffestiniog. Blodeuwedd is turned into an owl and banished to the night and Llew eventually dies to become the god of the Sun in some traditions.

This story not only ties with religious beliefs regarding the birth, death and re-birth of the sun from Celtic legends but is Christianized as the blameless life, death and rebirth of Jesus Christ.

Llew's power over the natural world is evident in the tale of the Cad Goddeu, or Battle of the Trees from the *Book of Taliesin*. His uncle Gwydion challenges Arawn the lord of the Otherworld and in the resulting battle Llew calls upon the oak, aspen, alder and flowers of the field to fight with them. As all of these trees are deciduous and the flowers perennial, the tale is evidently about life, death and resurrection as the Otherworld is the afterlife or death. The evergreen holly tree wins the battle which leads to the defeat of the forces of darkness. The natural plants and trees of Wales are well featured here so the next time one is wandering in springtime through a wooded glade filled with bluebells, recall the tale of Llew and the battle of the trees.

* * *

Perseus is one of the truly great constellations for the observer. Lying along the Milky Way as it does, the astronomer is guaranteed a feast of objects to

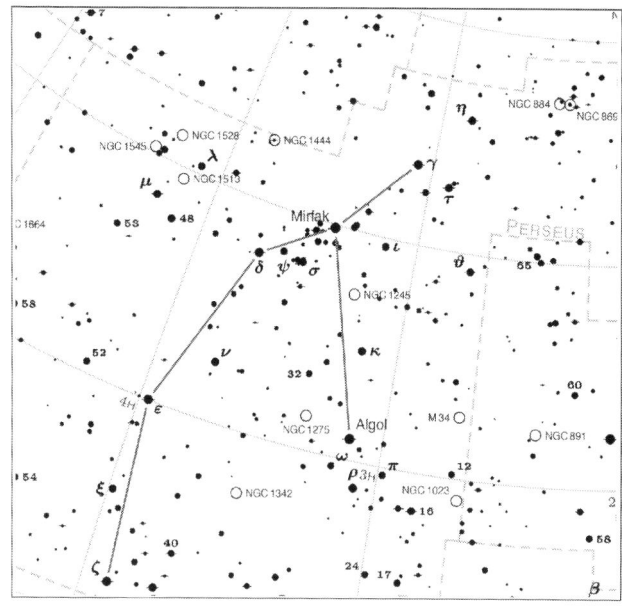

Perseus

delight and inspire, and this Perseus has in abundance. His figure sweeps along the galactic plane; sword and shield held high, feet firmly planted, it is easy to see the pattern of stars as none other than this heroic figure. Mirphak, the primary star of the constellation, is part of a large cluster of relatively nearby stars that are in turn part of a cluster or stellar association called the α Persei moving group, a wonderful sight in binoculars, as the companions are all bright O type stars numbering over 40 in a scattered little field.

Follow the upraised arm of Perseus along the Milky Way to a misty spot of diffuse light midway between Perseus and Cassiopeia, and you will be rewarded with the most exquisite star cluster in the whole heavens, or more appropriately, two clusters. This is the famous Sword Handle, a grouping of over 1000 stars in two small very rich clusters that are absolutely bursting with stars. In binoculars the two condensations are well seen, but in the eyepiece of a small telescope, the sight is awe-inspiring, the field is dripping with stardust and chains and patterns of all descriptions.

At the centre of one of the clusters a red giant sits prominently, ruling this ocean of starlight, proud in its majesty. The two clusters are NGC 869 and NGC 884, both very remote: some 7600 light years to NGC 869, and another

The Sword Handle Double Cluster

400 light years again for NGC 884. To shine with such luminosity to be naked eye objects, the stars of both clusters can only be supergiants of O and B type, whilst NGC 884 is the owner of the red giant mentioned above. Words fail to describe this tiny part of our galactic home; it is a veritable jewel box of stellar bodies in such profusion as to stagger the imagination. If this is the only thing you will ever see through a telescope, then it is no exaggeration to say you

Messier 34

will remember the sight for the rest of your life.

Messier must have thought them so obvious as to be not worth mentioning, an odd conclusion when he catalogued the Pleiades as M45, but he did not miss out on another showpiece cluster that now bears his stamp, M34. This cluster is approximately the size of the full moon, containing 50 stars of 6th-10th magnitude in a rich field that is easily visible in binoculars.

Before exploring other deep sky wonders of Perseus, why not examine β Persei, or Algol as it is also known. This is the prototype of a class of eclipsing variable stars and its variations are easy to follow with the naked eye. Usually, the star remains at mag 2.1, but at intervals of exactly 2.867 days it fades to mag 3.4, being eclipsed by a companion. John Goodricke, an eighteenth century astronomer who was a deaf mute, first explained this phenomenon. It is a pity that Goodricke died when he was only 21, as he had a genius for astronomy, and an originality of mind that could have led to important discoveries years ahead of their time. Algol can make compelling viewing, and many hardened variable star observers first cut their teeth on this wonderful object.

Following a line from Algol, back up to α Persei, you will encounter a rich field of stars that steadily agglomerate into a small but interesting cluster of 30 or so stars called NGC 1245. The field is worth exploring with binoculars, as several bright novae have occurred here in the past.

There is one particularly bright nebula in Perseus, but it is not generally easy to see with average equipment. This is the large 'California Nebulae', so called because it resembles a map of that American state. Ironically, it shows up well on a photograph, but is extremely difficult to view with the eye, although some observers claim to be able to see it with a pair of binoculars. Due to its low surface brightness, a pair of binoculars may show it well, whereas a telescope smacks of overkill.

Where the arm of Perseus curves around to hold his shield, there are several interesting star clusters worth observing. The finest of these is NGC 1528, a small cluster of 30 bright stars just above the curve of the arm. This group is again relatively distant at 2500 light years but it has a magnitude of 7.5, making it an easy object for binoculars, although the view through a telescope is more rewarding. Relatively close is another cluster, NGC 1545, a small but rich little asterism of 20 stars that is worth taking the trouble to find.

One of the most difficult objects in the whole constellation is, surprisingly, a Messier object, M76. This is a planetary nebula, often called the Little Dumb Bell as it resembles M27 in Vulpecula. It lies in a rather barren part of Perseus,

NGC 1528

close to the border with Andromeda, but is worth taking the trouble to find. It is not particularly difficult in a small telescope, although binocular watchers will have to put up with their disappointment. It shines at mag 10.5 and is a boxlike structure of hazy light, condensing around the edges. M76 is 2500 light years away and is visible in most telescopes despite being the 'faintest' Messier object in his catalogue.

Above the line of stars that make up the shoulders of Perseus are several rich clusters and some diffuse nebulae that provide a good night's hunting for the avid astronomer. The field here is particularly rich, as the Milky Way is unhindered by intervening dust lanes, and the Perseus arm of the Milky Way is relatively close to us. A pair of binoculars is a must to scan this lovely part of nature's garden of celestial delights. The area around the Sword Handle cluster merits attention as the Milky Way gleams with starry condensations at this point, and clusters flash into view.

Pisces Austrinus

The constellation of Pisces Austrinus, the 'Southern Fish' is fairly conspicuous, although it lies rather low in the south as seen from British latitudes. It appears to be a constellation of antiquity, as both the Babylonian and Persian empires recognised this constellation. It may also be a heavenly representation of the Phoenician god Dagon, although the constellation of Capricornus is usually attributed to this deity.

The Autumn Constellations

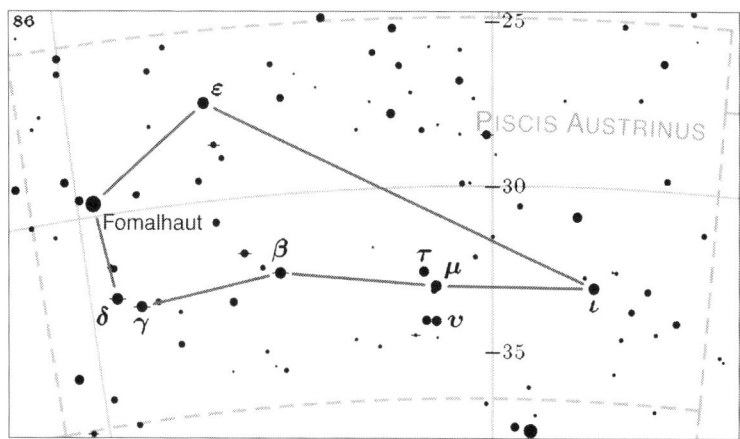

Pisces Austrinus

Pisces Austrinus is not mentioned in Welsh mythology though there are suggestions that the bright star Fomalhaut was part of the constellation of the Sea-cow or Capricorn as we would know it today. As a conspicuous star and the only first magnitude star very low on the horizon viewed from Wales, it may have played a part in autumn ceremonies but knowledge of such has since been lost.

Fomalhaut – its Arabic name meaning The Mouth of the Fish – is a lovely sight on an autumn evening, shining brilliantly with a lovely blue-white glow that scintillates through many colours of the rainbow due to its low elevation. It is a rather lonely star in a barren field, the only first magnitude star in the whole of the autumn sky, and sadly not remaining visible during this dismal season long enough for British observers to appreciate it.

Fomalhaut was one of the Persian 'Royal Stars', a group of four that apparently were responsible for the welfare of the royal house at different times of the year. The rest of this group are seasonal, they are Aldebaran in Taurus, Regulus in Leo and Antares in Scorpius. What astrological significance Fomalhaut would have is somewhat mysterious. It is the only Royal Star not to lie along the zodiac, but perhaps it was chosen due to its radiance, not its heavenly position. Fomalhaut is an A type star, similar to Sirius, lying 25 light years away, thus making it one of the closest bright stars to our solar neighbourhood.

Apart from Fomalhaut, the stars of Pisces Austrinus are not prominent objects; indeed, many of them lie below our southern horizon. There is a corresponding lack of deep sky objects too, which is not surprising considering the rarity of objects in its neighbouring constellation, Capricornus.

Yet there is one other star of interest within Pisces Austrinus, although observers in Britain will in all probability only be able to spot it at Culmination. This is the famous red dwarf star Lacaille 9352, a star with the fourth fastest proper motion in the sky. Lacaille 9352 lies about 12 light years away and has a total magnitude of 8, so it should be visible in good binoculars as a red glowing coal. To find this star, firstly locate Pi Pisces Austrinus then move your telescope or binoculars just one degree to the south. Lacaille 9352 should be the brightest star in this lowly, barren patch of sky.

Pisces
(The Salmon of Wisdom)

In Welsh myths the two fish of Pisces are mentioned as returning salmon, finding their way back from the oceans to spawn. Because salmon return to the same river as they were born in to breed and die, it was thought that salmon had remarkable wisdom and some tales mention that Taliesin, the ancient Welsh bard got his wisdom and powers of speech from eating the fat of a salmon which contained a sort of concentrated wisdom.

The salmon is a magical fish and appears in the *Mabinogion* in the tale of Culhwch and Olwen. In this story, a boar and a stag feature as do two birds and the Salmon of Wisdom, or Eog, and sources often associate the constellation as a pair of leaping salmon winding their way through rivers and streams jumping over obstacles in their quest to get home to spawn. To trace the hero Mabon, the knights Cai and Bedwyr are directed by the wise owl of Cwm Corlwyd in Snowdonia to the most knowledgeable creature of all, the Salmon of Llyn Llyw. The fish knew where Mabon was imprisoned and took the knights on his back down river to Gloucester where in the castle dungeons, they could hear the cries of Mabon and managed to free him. The animals of the tale are immortalised by Wales' famous poet R.S. Thomas in 'The Ancients of the World'.

Pisces must be one of the longest and faintest of all the constellations along the Ecliptic. However, once the outline can be traced it is relatively easy to recognize again. The constellation apparently consists of two fish, one, called the western fish, ends in a little oval of stars known as the Circlet of Pisces, whilst the eastern fish runs in straight line up the east hand side of the constellation of Pegasus, which lies directly to the north. Pisces is another constellation with an ancient history, once again it is one created by the Babylonians and passed down to future generations and cultures. The Egyptians thought of it as the two rivers, the White and the Blue Nile.

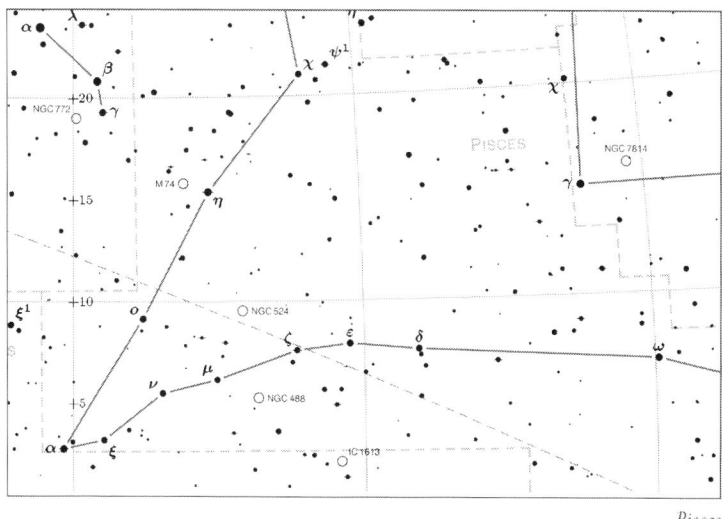

Pisces

* * *

Pisces contains a few objects of interest to those prepared to diligently sweep the heavens, although most of these objects are not at all bright, and are predominantly galaxies lying at remote distances.

α Piscium is an engaging double star, both of the components are apparently coloured, although observers cannot agree on the tone of each. This is a fourth magnitude star which is an A type luminary, which should be white in colour, but the famous observer T.W. Webb described α as the greenest star he had ever witnessed. Further observation by other astronomers have added such confusing shades as orange, yellow, yellow-brown or even blue!

With such dissimilar descriptions, it must come as no surprise that the fifth magnitude companion also appears to have various tints, ranging from white to red. α Piscium can be separated easily with a small telescope, although it is too close an object for binoculars, though the primary colour should show up well in such instruments. The observer will have to observe this object and come to their own conclusions, as the last of this argument has yet to be heard.

The finest deep sky object in Pisces is the beautiful face on Sc type spiral galaxy known to us as M74. It is a softly glowing halo of eleventh magnitude light, and must rank as the most difficult object to view in the whole of the Messier catalogue, even though there are intrinsically fainter objects. M74 may be seen under good conditions in a pair of giant binoculars, but a

Messier 74

telescope requires the use of averted vision to bring out the best in this object. It can be found midway between η and 105 Piscium as a blue mass of glowing gaseous light, with a hint of condensation towards the centre. M74 parallels our own Milky Way in size, it being only a little smaller than our home galaxy, and the observer can imagine from photographs of this object just how the Milky Way would look to an observer high above the galactic plane. M74 is 32 million light years away.

The only other galaxy available for scrutiny is NGC 488, at RA 01h 21m 48s Dec 05°15m; a lovely Sa spiral, lying 40 million light years away and glowing faintly at magnitude 11. It can be perceived in a small telescope as an elongated smudge of light, but is not visible in binoculars.

Pisces contains little else of note other than to observers with large telescopes that can enjoy the wonders of many more of these elusive galaxies buried in the depths of space.

Triangulum
(The Spearhead)

The constellation of Triangulum consists, as one would expect, of three stars lying in a rather uninteresting field to the south of Andromeda. It appears to be a constellation steeped in antiquity, although exactly what it is meant to represent is open to debate. It has been suggested that the Greeks thought of it as their capital letter Delta (Δ), but to the Celts this constellation was a white (iron) spearhead – possibly the spearhead that Goronwy used to kill Llew which dropped out of his body as his soul ascended into the heavens. It lies under his wife Blodeuwedd and fairly close to Llew in the sky so this interpretation may well be justified.

The three stars creating the triangular outline are fairly conspicuous, but there is very little of note to the deep sky observer except for two superb objects.

The first is a lovely double star called Iota Trianguli. This star is a coloured double, one of the finest little binaries for a small telescope. The separation is rather diminutive, only 3.8", so it could be a rather stiff test for small instruments, but the sight of this star, once found, will never be forgotten. Both components are apparently yellow stars, but the fainter companion has a noticeably bluish hue, the whole system looks like a miniature copy of Albireo in Cygnus. It may just be visible in giant binoculars, although the wide field of view sets these instruments at a disadvantage compared to a small telescope. Iota Trianguli is 305 light years distant.

The next object to ponder in this small constellation is actually one of the greatest objects in the autumn sky, after the distinguished Andromeda galaxy. It too is a galaxy, one known as M33, or the Pinwheel Galaxy as it has come to be known. It is an incredible object, lying almost face on to our view, shining at sixth magnitude with the combined light of billions of Suns.

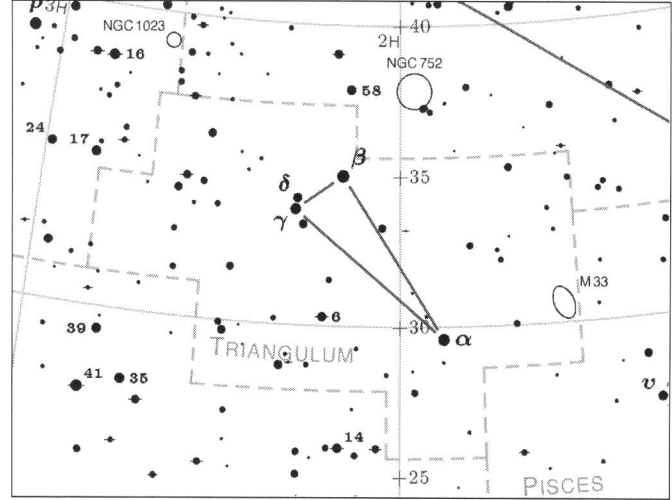

Triangulum

Messier 33

Unfortunately, due to its large angular size, it cannot be observed well through a small telescope, but it is a magical sight when seen through binoculars. It appears as a luminous greenish glow of light in hazy outline, with a barely discernable nucleus, which reveals itself only as a delicate brightening towards the centre of this misty glimmer of stars. Binoculars truly are the best instruments for this object, as you can find M33 relatively easily with them by taking a line from the tip of the triangle and moving upwards towards Andromeda. It is a little difficult to star hop to in a small telescope, but M33 can be seen as a blue-grey blur of light under clear conditions and it can be photographed relatively easily.

M33 is one of our local group galaxies, apparently lying a little further away than the Andromeda galaxy, at a distance of 3.3 million light years, and separated from it by over 500,000 light years. It contains the largest known ionised hydrogen region, or nebula, which is bright enough to be seen through a large amateur telescope. Designated NGC 604, it is over 1500 light years in diameter, compared to the paltry 26 light years for our own Orion nebulae M42, which to our eyes is a glorious cloud of light. What such a cloud would look like if it were as close to us as M42 can only be imagined; the night sky would be brighter than the full moon, and perhaps we would not have much opportunity to learn so much about the Universe if we lay within the blinding luminescence of such an object. Triangulum contains a few faint galaxies that are beyond the scope of most amateurs.

CHAPTER THREE
The Winter Constellations

Stars have ways I do not know,
Enormity that checks my thought,
yet on the loom of their fine glow
the fabric of my dreams is wrought.

I look into the stars, and one
after one, convictions die,
while more than I have lost is spun
delicately across the sky.

I look into the stars, and all
the fuming purposes life gives
pass, like mists of evening fall,
and all life never has been, lives.

— Jane Draper

Chapter Three – The Winter Constellations

The winter sky is undoubtedly the finest of the seasonal groups visible from Britain. The stars scintillate brightly in the frigid air, and the constellations are at their loveliest, dominated by the beautiful Orion, the key to the rest of the winter constellations. High overhead can be found the groups of Auriga and Perseus, lying to the east and above the shoulder of Orion is Gemini, whilst off the western shoulder of Orion can be found Taurus the bull. To the west of Orion is the dim and straggly trail of stars that mark the constellation of Eridanus. The sky below and to the east of Orion is filled with the magnificence of Canis Major, Lepus, Puppis, Monoceros and Canis Minor.

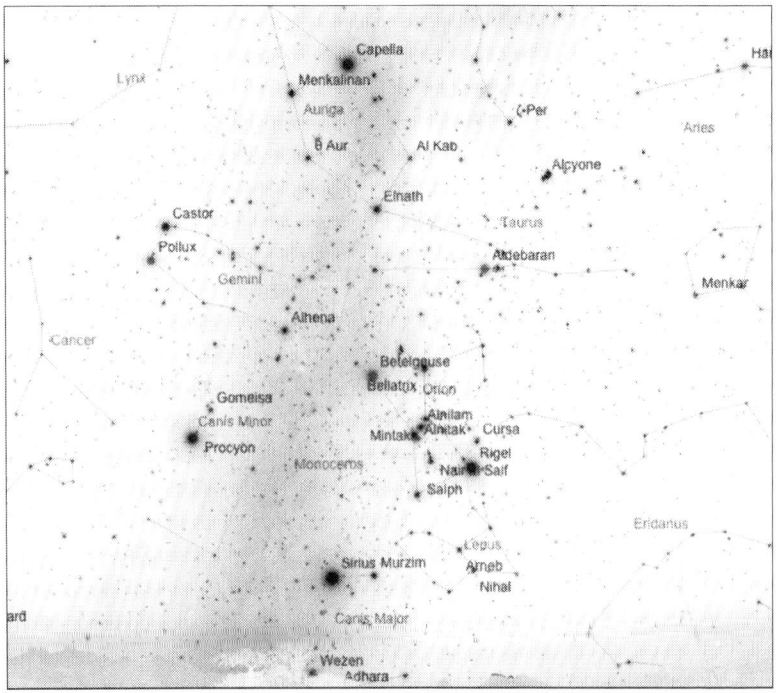

Winter Constellations

This dismal season is wet and cold in Britain, but the climate can sweep the dust and sediments out of the air, leaving the atmosphere as clear as crystal. In a high-pressure weather system, good observing conditions are guaranteed, although wrapping up very warm and taking a flask of something hot is essential. Such precautions cannot be emphasised enough, as many nights observing has been ruined by the observer returning home chilled to the bone within five minutes of starting their observing programme.

Gloves, scarf and hat are indispensable items that will prolong the session. A woollen hat can make all the difference between a good night outside or a miserable one warming up in bed! In addition, see if you can stand on a raised board or platform whist observing, as the chill of the ground will sap heat through the soles of your feet. If you do get cold, walk around, chat to any friends who may be with you, have a warm drink and take the occasional break.

A telescope must cool down to the temperature of its surroundings to work effectively, so rime ice and frost can make the scope become clammy, and can even 'burn' hands or fingers that linger too long on exposed metal parts, so touch the telescope as little as possible whilst actually observing.

If you are viewing the magic of the winter sky with binoculars, then the converse is true. It is much better if your binoculars are kept warm by body heat all the time they are not in use. Being a sealed unit they do not need to be kept at the temperature of the surrounding air to work well, but keeping them warm will avoid the moisture of your eyes condensing on the eyepieces When not in use, tuck them inside your coat to keep them warm. This may sound like simple common sense, but in forty years of observing I have seen many fall foul of taking the most elementary protection against the cold, with the result of numb toes and fingers, a shivering body, and worse of all the dreaded 'hot aches' of returning circulation. Although I have not heard of an amateur astronomer dying from exposure, there is always a first time! So, enjoy the glories of the winter sky warm, snug and protected from the elements.

The winter sky has a 'triangle' of stars that may assist the observer in finding other constellations. These bright beacons are Sirius in Canis Major, Procyon in Canis Minor and Betelguese in Orion. Look at the tint of each as they give a vivid indication of star colours. Procyon is yellow, Betelguese is orange-red and Sirius is a pure white. As they are the brightest stars in the winter sky, they should not be difficult to spot.

Low on the southern horizon during winter are some undistinguished stars that belong to the constellations of Columba the dove, Caelum the sculptor's tool and Pyxis the pump. Both Caelum and Columba are found below Lepus whilst Pyxis is on the horizon to the east of Puppis. There is nothing of note in any of these groups due to their very low altitude.

Distinctive Stars

The brightest (apparent) star in the sky is Sirius, spectral type A1V, in the constellation of Canis Major, the Great Dog. In fact, its magnitude is −1.46, reminding us that calling it a 'first magnitude' star is not strictly accurate, merely a throwback to earlier history when all stars in the first class of brightness were grouped together as first magnitude. There are several confusing appellations in astronomy that have been left over due to tradition; one of which being the common name of Sirius – the 'Dog Star' for reasons which should be obvious.

Sirius is the only star bright enough to show the colour changes caused by turbulence in our atmosphere. The show of these colours in the telescope is all the more amazing. But when you learn more about the nature of Sirius, its impressive appearance becomes still more interesting. It is the closest bright star visible from 51°N, and the sixth closest star system known in all the heavens; it is 8.6 light-years away. We say Sirius is 'near' for a star – yet light, the fastest thing in the universe, travelling at 186,000 miles per second, takes over 8 years to reach us from Sirius. Sirius is over 500,000 times farther from our Earth than the Sun is. And yet compared to most stars it is 'nearby'. Only one of the other dozen or so apparently brightest stars in its constellation, Canis Major is less than 200 light years from Earth. These other, more distant suns of the group are all of greater true brightness than Sirius.

Sirius has a famous white dwarf companion star. It is only magnitude 8 and unfortunately is too close to Sirius for easy observation. It was discovered by Alvan Clark, a telescope maker in the late nineteenth century, before it was truly appreciated what sort of body this was. As an aside, if you have identified this area, try moving your binoculars or telescope south from Sirius. Doing so should bring you to the relatively bright, attractive star cluster M41, which can be seen with the naked eye under good conditions.

Capella, spectral class G8III, in Auriga the Charioteer, passes high in the sky for mid-northern latitude viewers, and at 40°N is almost a north circumpolar star – a star whose apparent circling around the north celestial pole is never cut off by the horizon, and so never appears to set. The spectroscope proves that the point of light we see is actually produced by a close-together pair of stars, both of which are larger than the Sun, but rather similar to the Sun in spectral type. The two stars are too close together to see separately in the telescope, but it is possible to detect their yellowish hue. There is still considerable mystery about these suns, which are thought to revolve around each other in a nearly circular orbit and about 100 million km apart. One of the stars may have turbulent motion in its atmosphere.

Rigel, spectral class B8Ia in Orion appears imperceptibly fainter than Capella, but is much farther away and thus greater in true brightness. Rigel is in fact a blue giant star, and one of the most luminous stars in our part of the galaxy. At its distance of 900 light years it shines with a true luminosity of 60,000 times greater than our Sun. Of the first-magnitude stars, only Deneb in the summer constellation of Cygnus rivals it in true brightness.

Procyon, spectral type F5IV, is the lucida (star of greatest apparent brightness in a constellation) in Canis Minor, the Little Dog. It is, appropriately, considerably less bright in our sky than Sirius, the lucida of Canis Major the Big Dog. What is surprising is that Procyon happens to be not much farther from us than Sirius; thus, it is considerably less bright in reality. Procyon has a white dwarf companion like that of Sirius but it is smaller and dimmer.

Betelgeuse, spectral type M2Ia, shines brighter in our sky than any other red giant, and may be as large as any sun known. While its luminosity is far less than that of Rigel, there have been rare occasions when its apparent brightness rivaled or surpassed Rigel's. Betelgeuse is a variable star with a moderate range of brightness variation. The brightness changes are thought to be associated with pulsations in size due to stellar winds buffeting the tenuous envelope of gas in the star's outer envelope.

Direct observations of Betelgeuse by the Hubble space telescope reveal that the average diameter of the star is larger than the orbit of Jupiter. Betelgeuse probably has at least two companions, both much too close to it to see in the telescope. In fact, one is so close that it may orbit within the dimly glowing and incredibly tenuous outer layers of Betelgeuse – a star within a star! Betelgeuse was the first star to have features identified on its surface, corresponding to sunspots on our own Sun, but vastly larger. Study the colour of Betelgeuse and note how that colour may vary with its changes in brightness, although this may be a long-term project.

Aldebaran, spectral class K5III, appears to be the brightest star in the Hyades star cluster, but is actually about half the distance from us compared to the other cluster members, and unrelated to them in space. Aldebaran (the eye of Taurus the Bull) is almost as distant and as luminous as the spring star Regulus at 65 light years, but Aldebaran is a far cooler star (note the orangish hue of Aldebaran) and thus much larger than Regulus and the Sun – it is in fact about 20 times the diameter of the Sun. Aldebaran has a large space velocity and is receding from us rapidly.

Pollux (spectral type K0III) and Castor (spectral type A1V) lie about 4° apart, the twin bright lights of Gemini the Twins. Their true brightness seem to be almost identical, but Castor is a little farther away and thus appears somewhat dimmer. Compare the colours of Castor and Pollux in the telescope, noting that Pollux is the redder of the pair and thus is considerably larger. A profound difference between Pollux and Castor, however, is that Pollux seems to be a single star, whereas the Castor system contains at least six stars. A small telescope at a fairly high power shows Castor as two bright and beautiful points. A redder, dim companion is at a considerable distance away.

Adhara, (spectral type B2II) located in southern Canis Major, is (like Castor) just a little too faint to have been considered a first-magnitude star. It actually appears a little brighter than the far more famous Castor. It is a B-type star and its luminosity is very great, about half that of Rigel.

Lying along a north-south line from horizon to Horizon is the winter Milky Way, part of our galactic home. Explore it with binoculars as there is much to see, but perhaps a brief introduction to this river of light would be instructive.

The Milky Way in History and Culture

The Milky Way as understood by modern astronomers has enjoyed a long evolution in the history of ideas. The Egyptians considered the starry band across the sky to be a pool of cow's milk, which they associated with the goddess Bat. To the Hindus of India the Milky Way forms the belly of a dolphin that assists the planets in their path across the sky. It is called 'Asakaganga', meaning the Ganges of the sky.

To the Welsh, the celestial river is Sarn Gwydion, the road of Gwydion, the hero who went into the heavens to fetch the soul of Llew and return him to life. In China it was seen as a cowherd and weaving girl, their love separated by the Milky Way, a story that is today a central part of the Qixi festival, a sort of Chinese Valentine's day.

To the Maori the Milky Way is the canoe of their god Tama Rereti; to the Aborigines it was the Wodliparri, the Sky River, along which dwelled people and great monsters. In the past, tribes segregated the Milky Way into two parts by utilizing the Cygnus Rift, making two distinct camps in the sky. The Kung Bushmen of the Kalahari thought the Milky Way was the 'backbone of night', signifying the encompassing of their world as the inside of a great animal. The Khoisan people of Africa have an oral tradition that states long ago there were no stars and the night was pitch black. A girl, who was lonely and wanted to visit other people, threw the embers from a fire into the sky to light her road and created the Milky Way.

In the world of the ancient Greeks, the Milky Way was produced by the milk of the goddess Hera, spraying across the sky after suckling Hercules. The Greek word *Galaxias* reflects this legend, which was later taken up by the Romans who named the river of stars *Via Lactia* from which we get our English translation Milky Way. The word galaxy is now applied to all such large systems of stars in the cosmos but it is easy to judge the importance of this celestial marvel to ancient peoples and cultures.

Some of the first people to rationally determine the Milky Way to be a collection of stars were Anaxagorus, in the fifth century BCE and Democritus, in the fourth century BCE, both of whom made the statement that the Milky Way was a collection of distant stars. The Islamic astronomers Al Biruni and Al Tusi wrote that the Milky Way was made of innumerable small and distant stars, here and there collected into groups, but at such distances to give the appearance of nebulous clouds.

The Italian astronomer Galileo Galilei made the final determination by using the telescope in 1610 and found that the Milky Way was indeed a vast collection of faint and numberless stars spread in a disk-like shape. The concentration of stars led the philosopher Immanuel Kant to predict that the Milky Way was in fact an 'island universe', a self-contained body of stars which had counterparts spread throughout the cosmos. The eighteenth century astronomer William Herschel demonstrated the disk-like nature of the Milky Way, and was the first to determine the apparent movement of the Sun with respect to the disk. The confirmation of the Milky Way as an island universe or galaxy was finally resolved with the work of Vesto Slipher and Edwin Hubble in the twentieth century.

In 1913 the astronomer Harlow Shapley finally determined the sun's position within our galaxy. Although he used the relative positions of globular clusters in the galactic disk, there is a little appreciated fact that gives us a clue about our position in the Milky Way. In winter, the Milky Way is relatively dim in comparison to the splendour of the summer Milky Way, yet its stars are bright because they lie on a close by spiral arm, the Orion arm that also contains the Sun. In winter, we are looking out through the Orion arm into intergalactic space – the stars may be brighter due to their proximity, but the Milky Way is sparse as it is less dense in our direction of sight. Conversely, the summer Milky Way is brighter and more condensed as lying along our line of sight is contained the whole of our galaxy, with its centre in the southern constellation of Sagittarius. In autumn and spring, the Milky Way is lying all around the horizon and so we look out into intergalactic space over the north galactic pole in spring, and the south galactic pole in autumn. As the Milky Way is a disc-shaped galaxy, the stars above the plane of the disk are few in number and so the spring and autumn constellations have few bright stars. This accident of our position in the galaxy accounts for the diversity of seasonal stars we perceive.

The Winter Milky Way

Although not as splendid as the summer Milky Way, the winter portion of our galactic home still holds many surprises. Stretching from the northwestern to the southeastern horizon, the Milky Way is not as obvious as it is in summer, yet the stars along its path are brighter and glitter beautifully in the frosty air. The observer is looking through the Orion-Cygnus and Perseus spiral arms; the outer arms of our Milky Way in winter, so the grandeur of the galactic centre or the combined light of many stars in the spiral arms is not seen. Nevertheless, one is looking at a great oddity here, a feature known as the Gould Belt.

First identified by the astronomer Benjamin Gould in 1879 the belt contains the majority of stars that form all the constellations from Puppis through to Perseus: almost the entire sweep of the winter sky. The belt is about 3000 light years in extent and inclined to the Milky Way by 20 degrees. Recent investigations suggest that this formation is due to the collision of a dark matter cloud with the arms of our galaxy about 30 million years ago, creating many of the O and B class stars that illuminate our winter sky.

From the constellation of Perseus, note the double cluster h and χ Persei

Winter Milky Way

and their surroundings where larger clusters such as Melotte 15, NGC 957 and NGC 744 flank the brighter concentrations of the twin gems, studding the area with diamond light. The double cluster is more distant than our local Orion arm at about 7000 light years away and the objects are separated by about 400 light years, creating a beautiful line of sight effect. Sweeping across the scatter of stars in the Alpha Persei moving group and noting the starlight there one can move into the constellation of Auriga and observe the predominance of stars around 18 and 19 Aurigae and the milkyness of IC 405, the Flaming Star Nebula, which is not seen to great effect but hints at its presence.

Darting down the Milky Way in a line from here are the three great Messier clusters M38, M36 and M37, which are always worth exploring, whilst terminating this part of the Milky Way is the wonderful star cluster M35 just over the border in Gemini. As the galaxy knows no boundaries, it's a straight run through all four of these magnificent objects. Try to spot NGC 2158 in the same field as M35, a dimmer and more concentrated cluster.

The observer can then detour slightly to examine the Gould Belt in more detail by exploring the constellation of Orion with its seven bright stars; the only one that is not a belt member is the red supergiant Betelgeuse, the others are at similar distances and age. With binoculars, check out the cluster Collinder 70 which winds in intricate patterns around the three stars of Orion's belt. M42 and M43 are obvious targets, making a bright splash of light in the 'sword'; they are in fact bright parts of an enormous molecular cloud that stretches all the way into Auriga. The nebulae along the length of this cloud are just the tips of a dark iceberg hidden amongst the stars of the galaxy here.

Crossing east to the Milky Way from Orion the observer can pick out NGC 2264, the Christmas Tree cluster and the large amorphous 'Rosette nebula' with its cluster NGC 2244 as a block of rectangular stardust at the centre. Heading southward through Monoceros the observer encounters the constellation of Puppis with its bright star clusters M46, M47 and M93 before crossing into Canis Major to examine the stars of the dog's hind-quarters, especially τ Canis Majoris and its concentrated little cluster NGC 2362. This lovely object will be hard to distinguish in binoculars, though a small telescope reveals the full glory of the 'Northern jewel box'.

Here the Milky Way fades into indistinctness as it meets the southern horizon from the UK, but the journey around our galactic home has been dazzling and very worthwhile. If one wishes to attempt something a little different, take a relaxing ride along the starfields of our galactic home. The Milky Way and the stars described above are a good introduction to discovering the following series of constellations that are designated here again in alphabetical order.

Auriga
(The Yoke or Bondsman)

The constellation of Auriga, the Charioteer is one of the Greek astronomer Ptolemy's original groups, easily discernable to the eye. The story of how Auriga came to be placed in the sky is obscure, but it is thought that this figure commemorates the establishment of the wheeled car or chariot, hence the name, though the driver is credited as Erichtonius, the lame son of Hephaestus and Artemis who invented the chariot as an easier form of locomotion.

In Welsh legend this constellation represents the yoke borne by Taurus the bull as it ploughed the sky, dragging the constellation of Ursa Major (the plough) and held in line by Hu Gadarn. In some translations of ancient tales, it also represents a bondsman who has the traces from the plough over his shoulders and is urging the beast onward by patting its back with a stick.

* * *

Auriga is defined by a quad of four bright stars rather like a child's kite in outline. The primary star of this constellation is the yellow giant Capella, a beacon that captures the attention on a winter evening. As Capella lies at 50 degrees declination, it passes through the zenith as seen from Britain, thus enabling us to view the whole constellation to advantage, high in the night sky. Capella is a close star, 42 light years away and is 160 times brighter than the Sun.

Capella are three stars forming a small triangle. The name Capella is translated 'little she-goat', and so this group is commonly referred to as the Heidi or 'kids'. One of the kids is extremely interesting. Auriga is a yellow supergiant star that is a special type of eclipsing binary. The eclipses occur every 27 years, the last being in 2009-11, and a single eclipse lasts for over a year! It is not known what kind of body is responsible for such a long eclipse; many think that it is another supergiant star surrounded by enormous shells of dust and gas. During the last eclipse, observers all over the world were encouraged to keep an eye on the star and report the changes in magnitude as the light slowly

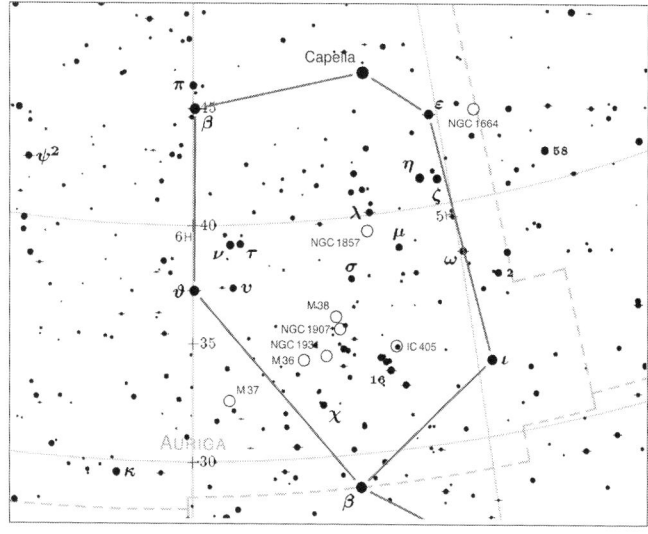

Auriga

dimmed. What they found was that as the eclipse progressed, the primary star flickered slowly, giving rise to the theory that clouds of dust and gas rather than shells of gas are accountable for this phenomena, the light of the primary shining through the gaps in the clouds, causing the scintillation seen from Earth. Auriga could be part of a binary system that is either very young and still collapsing to form a main sequence star, or more likely, its companion is surrounded by a disk of material. Several stars of this type are known.

β Pictoris, for example, has a disk of gas in its equatorial plane, and so does the lovely blue white Vega. It is conceivable that ε Auriga is a line of sight effect, this disk occulting the primary star, with an orbital period of 27 years.

Auriga lies in a very rich part of the Milky Way, and is itself rich in clusters and other deep sky objects. A famous trio of clusters can be found strung in a line through the middle of the constellation, making a remarkable sight in a pair of binoculars. The clusters are M36, M37 and M38, three striking gems that delight the observer. M38 is the most westerly cluster, and can be found close to the point of a line of 5th magnitude stars in the lower part of Auriga. This cluster contains over 100 stars, and is very rich and compressed, but what makes it such an outstanding object is that it is shaped into an easily distinguishable 'cross' of 8th magnitude stars that fill the eyepiece of a telescope. The cluster twinkles with unresolved pinpricks of light as the Milky Way provides a fine backdrop for this lovely object. Close by M38 is an

Messier 38

intriguing little cluster of stars, much more remote, almost merging with the Milky Way. It is NGC 1907, around 3500 light years away. Its blue giant stars shine at approximately 10th magnitude. Observers using a small telescope should be able to pick it up quite easily.

Messier 36

The next object is the wonderful M36, not as striking an object as M38, but a nice compressed group nevertheless. M36 contains around 60 stars of mag 9. In binoculars it looks like a hazy patch of gaseous light, although giant binoculars will reveal its true form. Nearby is the reflection nebulae NGC 1931 at RA: 05h 31m 24s Dec: 34°15m, visible as a compressed white patch of filmy gas amongst the stars of the Milky Way to the southwest of M36.

The showpiece of Auriga is the final object of the trio, M37. This cluster lies at a similar distance to the others at approximately 1000 light years, but contains the most amazing accumulation of stars. Over 200 brilliant suns dominate this cluster, compressed into a field of view that is rather small in comparison to the clusters M38 and M36. It is extremely rich: the first time most observers see this wondrous object they are astonished at the lavishness of the cluster. Stars seem to radiate out of the eyepiece at you, and the longer you look the more stars become apparent. This cluster is one of the most inspiring objects in the heavens, and one that will remain in your memory long after your observing session is over.

There are several other objects in this region, which is shot through with diffuse nebulae. Unfortunately, many of these are very dim objects even in a large telescope. The best of these is the nebula I.C. 417 close by the star AE Auriga. This is a beautiful, deep-red emission nebulae, whilst AE Auriga is an

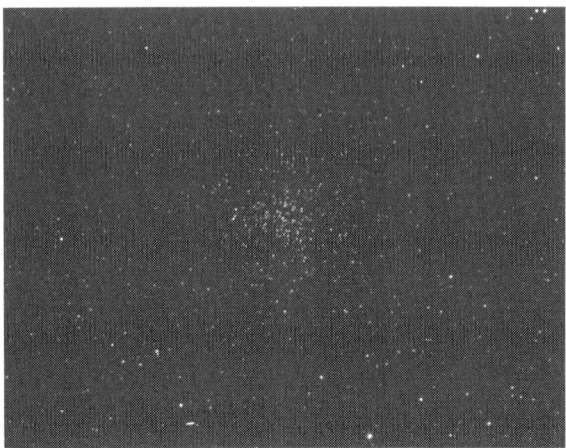
Messier 37

interesting star in itself; it appears to be rushing through space at high velocity, much faster than rivals at comparable distances. It is believed that AE Auriga was part of a large binary system that underwent a supernova explosion; the companion star was ejected from the system as a result of its high orbital velocity, and has become a 'runaway' star. There are a number of known of this type, so AE Auriga is not exactly a rarity, but is of interest nevertheless. The Milky Way runs through Auriga and it is worth exploring with binoculars as subtle dark clouds and knotty concentrations of star clusters abound in this area.

Cancer

The constellation of Cancer, the Crab, is an undefined grouping of faint stars in an area between Gemini and Leo. It is an ancient constellation, although why it should have been given such prominence is not known, perhaps early astrologers created the asterism to fit into theories that the monthly movement of the Sun through various constellations had some bearing on Earthly life. Legends abound regarding this little creature; the Greeks thought the crab came to the assistance of Zeus when he wrestled with his brother Poseidon for mastery of the Earth. Poseidon killed the crab, whereupon a victorious Zeus rewarded it with a place in the heavens. It also features in the myth of Orion and that of Hercules as an unfortunate creature sent by Hera to antagonize the heroes, but was killed and placed in the sky.

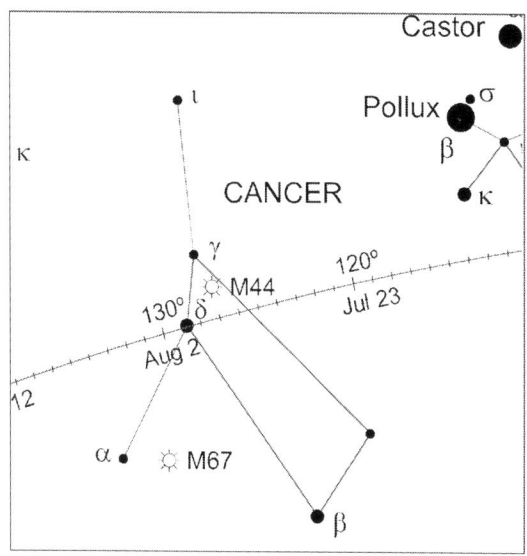

Cancer

There are very few associations in Welsh tales to the constellation, it possibly being a part of another group or standing alone but not being seen as crab. The only references in Welsh literature to crabs refer to the crab-apple rather than the sea creature, as the Druids thought of the apple and crab apple trees as being useful in ceremonies and a semi-divine gift. In some tales there is a vague reference to Dylan, the firstborn son of Arianrhod, who became a sea creature that was eventually killed by mistake by his uncle Gwyddon, or Gwydion in some sources. Dylan was renown for his swimming ability so it is more than likely that the tale refers to a fish rather than a crab as crabs are better known as walkers rather than swimmers! Nevertheless, the crab was a source of food along the Welsh coasts as it still is today, so perhaps the inclusion of this interesting creature has some merit in the old tales even though its proper meaning may be lost.

* * *

Cancer contains two of the most absorbing star clusters in the heavens; one of which is a treat for binocular observers. Disappointingly, Cancer has few other objects of note despite it astrological prominence.

The cluster that most casual observers will be familiar with is the wonderful M44, the Praesepe, or Beehive, or Manger as it is called. Praesepe is one of the closest clusters to the Earth [lying 525 light years away] and is visible as a small cloudy patch of light to the naked eye. Its appearance had led it be remarked upon by every ancient civilization, being known as the 'cloudy one' or 'misty one' by famous astronomers of the past. This may be a reference to the fact that its appearance was used to forecast weather, the poet Aratus, writing in the *Prognostica:*

> A murky manger with both stars
> shining unaltered is a sign of rain

Praesepe contains over 200 recognised members, but for the most part these are faint dwarf like stars that are impossible to see in binoculars. The stars that give Praesepe its lovely visual appearance are all white and blue O and B type suns shining with luminosities of up to 150 times that of our Sun.

Messier 44

Praesepe is brilliant in binoculars, as they define the cluster rather well and show the central condensation of stars plus the doubles at the core of the cluster. A telescope will not show Praesepe to great effect because its high magnifying power tends to look 'through' the cluster rather than at it. Telescopes will however resolve the central binaries well. The cluster core has a diameter of 13 light years, and contains over 60 stars that can be seen with a modest telescope.

Praesepe is a curious object in that it shares some common qualities with the Hyades in Taurus. Studies of the clusters show that their common proper motions are very similar. It is possible that they were born around the same time and place in our galaxy and have become slowly separated, although this assumption is total conjecture. However, the cluster is a real showpiece object, one of the kind that make a cold winter night in the darkness worth the effort. Another cluster of particular interest is M67, lying east of α Cancri This is a lovely compressed gathering of stars numbering over 500 suns in total. However, to a binocular observer it looks like a hazy smudge of 7th magnitude light, whilst telescopes will reveal the 70 or so central bright members of this awesome cluster. The main attraction of M67 is that it appears to be the oldest star cluster yet found in our galaxy – at least 4 billion years old.

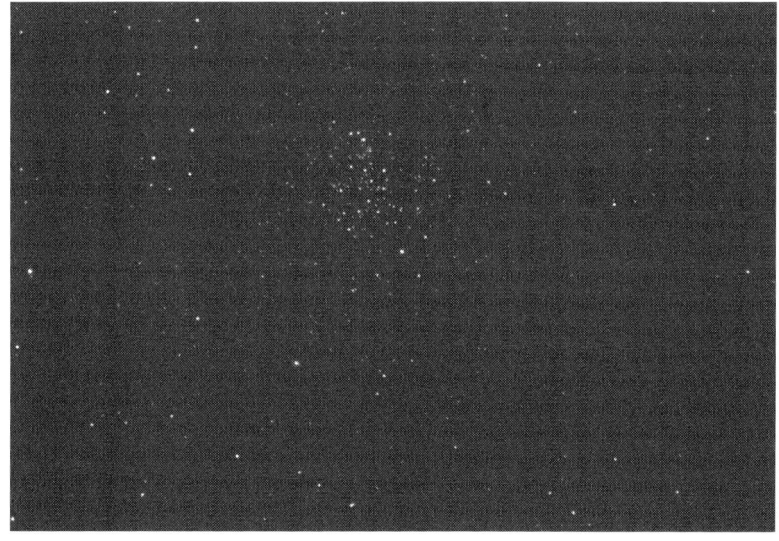

Messier 67

Most of the stars within M67 are K type giants evolving away from, or already evolved from the main sequence. In addition, the luminosity of these giants appears to be far below that of usual stars of this type. This could be due to the fact that we are looking at population II stars, very old halo objects rather than disc population I stars. Therefore, the chemical compositions of the stars are slightly different; these being deficient in metals compared to stars such as our Sun, and this may be the reason for their low luminosity.

M67 lies at a great distance from us, 2600 light years, and some 1500 light years above the galactic plane. The cluster appears more like the globular clusters in composition than the normal galactic clusters. NGC 188 in Cepheus has a similar life history and composition. M67 is not difficult to pick out with a telescope as it lies close to some 5th magnitude stars that act as stepping stones to it from α Cancri

Another object worth viewing in this small constellation is the star Zeta Cancri, which is a multiple system. However, most observers will only spot the one companion as the closer companion to Zeta is not well resolved, lying very close to the primary and visible only in very large telescopes.

Cancer can become a fascinating constellation due to the fact that it lies along the Zodiac or the plane of the ecliptic. The bright planets can then pass close by such famous objects such as M44 making an interesting contrast and a good target for photographers. For example, in 2002 Jupiter passed very close to this cluster, presenting a pleasing photographic opportunity.

Canis Major
(The Story of Gelert and Drudwyn)

Canis Major, the hunting dog, is an ancient constellation that has some wonderful associations. One hunting dog linked with this brilliant constellation retells one of the epic tales of old Wales, that of Gelert the hound of Llywelyn the Great. In this legend, Llywelyn returns from hunting to find a scene of devastation in his rooms with furniture shattered, clothing everywere, his baby son missing, the cradle overturned, and the dog Gelert slinking around with a blood-smeared mouth. Believing the dog savaged the child and wrought such destruction, Llywelyn draws his sword and kills poor Gelert, his faithful hound. After the dog's dying yelps had ceased, Llywelyn heard the cries of the baby, completely unharmed under the overturned cradle, along with the huge body of a dead wolf which had tried to attack the baby and been fought and killed by brave Gelert. Llywelyn was so overcome with remorse that he buried the dog with great ceremony, his grave still visible in Beddgelert. It is said that after that day, Llywelyn never smiled again, though faithful Gelert took his place in the sky.

As Mabon the hunter can be identified with the constellation of Orion, then in addition to the tale of Gelert the group of Canis Major can share an association with the hunting dog Drudwyn from the *Mabinogion*. Drudwyn was the only dog that could hunt down the Twrch Trwyth, the woodland boar. However, to use the dog, Mabon had to be freed from imprisonment by Culhwch as he was the only person who could control the magical hunting hound. With the help of Cai and Bedwr, Culhwch does so, and the boar is hunted down and presented to the giant Ysbaddadan so that Culhwch can marry his daughter Olwen. This magical hound then disappears, perhaps to take up its place beside its master in the sky.

Lady Charlotte Guest, the translator and publicist of the *Mabinogion* in the late nineteenth century, drew attention to Carn Gafallt in the Elan Valley as a place where the Twrch Trwyth was hunted by Mabon in the Arthurian legend of Culhwch and Olwen. Drawing on the tales of the ninth century monk Nennius who records in his *Historia Britonum* that the dog Drudwyn made a large mark on a rock on this mountain during the hunt, Lady Guest sent a man to the top of the hill to see if the indentation of the dog's paw was still visible and he reported that it was; though his scepticism also shone through as he stated: "The stone is near two feet in length, and not quite a foot wide, and such as a man might, without any great exertion, carry away in his hands. On the one side is an oval indentation, rounded at the bottom, nearly four inches long by three wide, about two inches deep, and altogether presenting such an appearance as might, without any great strain of imagination, be thought to resemble the print of a dog's foot… As the stone is a species of conglomerate, it is possible that some unimaginative geologist may persist in maintaining that this footprint is nothing more than the cavity, left by the removal of a rounded pebble, which was once embedded in the stone."

* * *

Canis Major is immediately recognizable due to the presence of the brightest star in the sky, Sirius or the Dog Star. In outline, the constellation looks nothing like man's best friend, but is a rough trapezoid of stars, with Delta, Epsilon and Eta making up a prominent triangle at the eastern bottom corner of the group. Sirius is one of the closest stars to Earth, situated 8.5 light years away, and shines with about 23 times the luminosity of our Sun, and is just over twice its size in diameter.

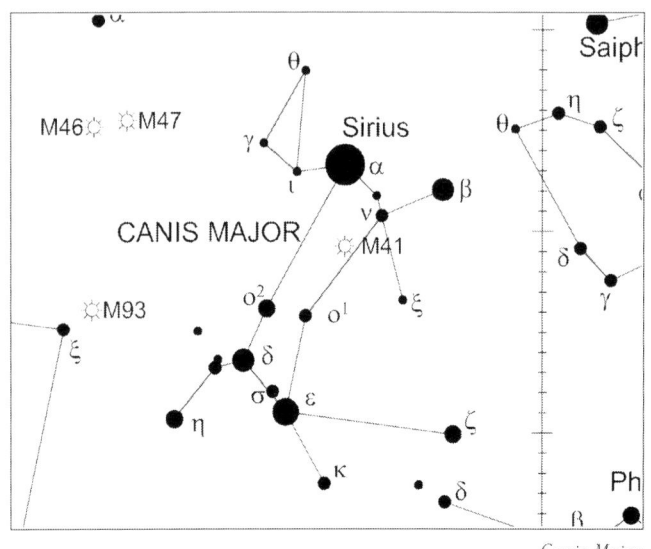

Canis Major

An interesting feature of Sirius is that it has a companion which was the first white dwarf star ever proposed upon collective observational evidence. Studies of the star by the German astronomer Friedrich Bessel, found that Sirius wobbled in it path of proper motion across the sky as measured by observers over a period of centuries, and he put forward the notion that an unseen companion was responsible for this reeling motion, or precession as it is called.

Bessel was vindicated early in the twentieth century when a faint star very close to Sirius was discovered by the great telescope builder Alvan Clark, testing one of his latest creations. This companion was dubbed 'the pup' or Sirius B, and may be seen under exceptional circumstances by observers with above average telescopes, although it is usually overshadowed by the luminosity of Sirius. The physical size of the star has been a matter of debate for some years, but from well established stellar theory it can be deduced that the pup is slightly larger than the Earth in diameter, yet contains as much mass as our Sun! The material of Sirius B must be extremely dense; a matchbox full of white dwarf matter would weigh many tons.

Sirius is also interesting in that it belongs to a very close cluster of stars known as the Ursa Major Stream, an aggregation of over 100 stars in an elliptical area of space about 150 light years in area. It is not the brightest star in this association however, merely one of the closest. How close Sirius lies is a matter of concern to some astronomers, as it is likely that it will evolve more rapidly than our Sun, turning into a red giant in the next 20 million years. The proximity of the pup to the main star will make the Sirius system a candidate for a type Ia supernova at some remote date in the future.

The outstanding deep sky object in Canis Major is without doubt M41, a large star cluster lying below Sirius. It is easily visible in binoculars as a starry smudge of 6th magnitude light, and some stars may even be resolved with such modest instruments. In the eyepiece of a small telescope M41 is a spectacular sight, the field ablaze with blue-white stars, condensing slightly towards the centre of the cluster. Although M41 only contains 25 to 30 stars, the impression one gets is that there are far more in the field of view. This is not surprising, since Canis Major borders the Milky Way, so a telescope will reveal more stars than properly belong to the cluster. M41 can even be seen with the naked eye under good conditions.

Messier 41

Travelling down the eastern border of the constellation with a pair of binoculars you will see a small cluster of stars surrounding the star τ Canis Majoris. This is the star cluster NGC 2362, a fine sight in such an instrument. In a small telescope, the field is filled with glittering gems, as NGC 2362 contains over 40 stars in a compact area lying over 4500 light years away in the Milky Way. NGC 2362 is of interest to modern astrophysicists as it appears to be one of the youngest star clusters known. All its stars are O and B type super giants, but as yet no nebulosity has been detected around them that compares with that of the Pleiades.

Just to the southwest of this cluster lies another interesting array, NGC 2354, easily found with a telescope just off the star δ Canis Majoris. This cluster contains around 60 stars in a rich group, but for the most part they are a relatively faint 9th to 13th magnitude, and therefore may not be visible in an average pair of binoculars. Another arresting batch for a small telescope is NGC 2360, not far to the east of γ Canis Majoris, marked by a fifth magnitude star on its western border. NGC 2360 contains 50 stars in a rich field, most of which are around 9th-10th magnitude, with a little chain of stars visible near the centre. The Milky Way provides a breathtaking backdrop to this cluster and many stars will be visible in the eyepiece.

Any good star atlas or app will show several other clusters of note, plus a number of galaxies that lie within the confines of the constellation, although these objects are far too faint for binocular or small telescope viewers. However, sweeping along the Milky Way is a pleasant experience, as some of the brightest portions of the winter Milky Way are found in Canis Major.

Canis Minor
(The Black Dog)

This is a small constellation, easily perceived as its brightest star is a first magnitude luminary that immediately attracts attention and is named Procyon, meaning 'before the dog' a reference to its rising before Canis Major. Canis Minor is an unremarkable constellation that has its roots in antiquity. It was one of Ptolemy's original constellations, and has always been referred to as a dog since time immemorial, although strangely, the Greeks recognised both Canis Minor and Major as one constellation. In later ages they thought of it as separate from Canis Major and named it the 'puppy'.

To the Welsh, this dog was the Gwyllgi, a dark magical hound that appears as a terrifying apparition of a Black Wolf with baleful breath and blazing red eyes. Wolves roamed the Welsh countryside for centuries until they were hunted to extinction, the last one being killed near Ystradfellte at a farm still known as Blaen Sawdde around 1760. In old tales the dog is often referred to as The Dog of Darkness or The Black Hound of Destiny as meeting the creature was meant to foretell the death of the person who encountered it.

Apparently, there have been many sighting of this beast in the north east of Wales especially near the Nant y Garth pass at Llandegla in Denbighshire.

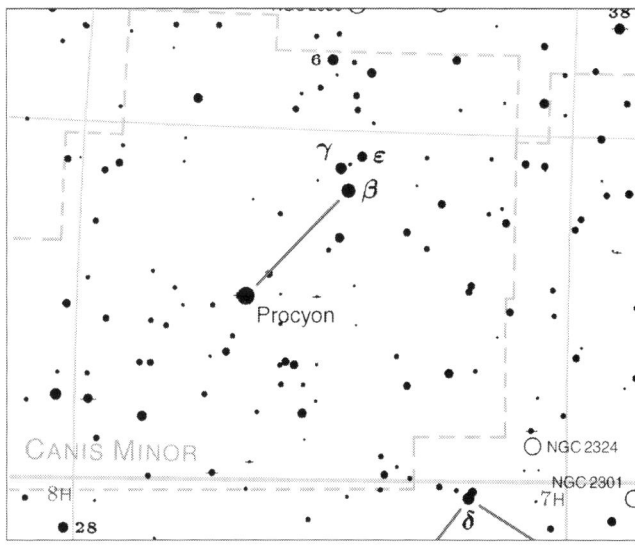

Canis Minor

It has even been spotted as far away as Marchwiel in Wrexham and it is said that this dark beast still roams the dark roads of Wales looking for victims. It has a lot in common with the Black Dog of English folklore too, as encountering this creature was meant to bring death or bad luck. Many pubs on both sides of the border are known as the Black Dog in homage to this hound of the night.

* * *

Canis Minor is the second hunting companion of Actaeon and is named Meara; it lies on the border of Monoceros to the west and Gemini to the north. It can be seen as an asterism of just two stars that lies in a northwest line tying the pair together. How this constellation ever came to be identified with a dog is one of the minor mysteries of the heavens, yet all ancient cultures refer to it as nothing else, with the exception of the Chinese, who thought of this constellation as a river, possibly because of its proximity to the Milky Way.

    The primary star of this pair is the beautiful Procyon, one of the brightest stars in the northern sky and the eighth brightest in the entire sky. It is also the fifth nearest to Earth of all the bright stars as it is positioned around 11 light years away, and has a luminosity of six times that of the Sun and is twice its size in diameter. Procyon is also commonly known as the 'little dog star'. Its rising before the Sun on the 19th July is thought to initiate the balmy 'dog days' of summer.

The other bright star of Canis Minor, β, is known as Gomeisa, but is unfortunately a rather boring object, lacking the saving grace of even being a double star. Despite this negative aspect, Gomeisa is positioned at a distance of 210 light years, and shines with the luminosity of 230 suns.

Surrounding the two major stars of this little constellation are several coloured doubles. Sweeping carefully with binoculars may reveal some of these little jewels, or at least the presence of stars of all colours in glorious profusion as Canis Minor lies quite close to the Milky Way. Scanning with a telescope will reveal many such stars to whet the appetite of those observers who enjoy such scenic wonders. If one enjoys this kind of activity, then the Webb Deep Sky Society's book of double stars is highly recommended, as it will uncover many such objects to delight and inspire during the long nights observing.

Eridanus
(The Great Rivers of Wales)

The constellation of Eridanus must be one of the most confusing from an amateur point of view. Eridanus – The River – is not too difficult to pick out, in fact the star β Eridani lies fairly close to β Orionis, the brilliant Rigel, whilst the rest of the constellation meanders in a line of faint stars south and westwards from this point, ending with the first magnitude star Achernar, which never rises in British climes.

In classical mythology, Eridanus is thought to be the river that the Earth disgorged to drown the rash youngster Phaeton, who thought he could control the chariot of the Sun that belonged to his father Apollo. Phaeton lost control of the chariot, and to prevent harm to the inhabitants of the Earth, the river claimed him, but not before he had burned the inhabitants of the continent of Africa and thus mythically accounted for their skin colour.

In local Breconshire mythology Eridanus can be identified with the twin waterfalls on the river Pyrddin. Sgwd Gwladys and Sgwd Einion. Gwladys, one of the twenty-four daughters of the King Brychan, had fallen in love with the shepherd Einion but her father the King of Breconshire did not desire his daughter to marry beneath her station. Consumed with grief at her lost love she retired to the waterfall and her tears kept the falls flowing even in summer. Her shepherd love also pined away and the two lovers are now joined in the winding Pyrddin with Sgwd Einion at the head of the valley and Sgwd Gwladys midway down the twisting river.

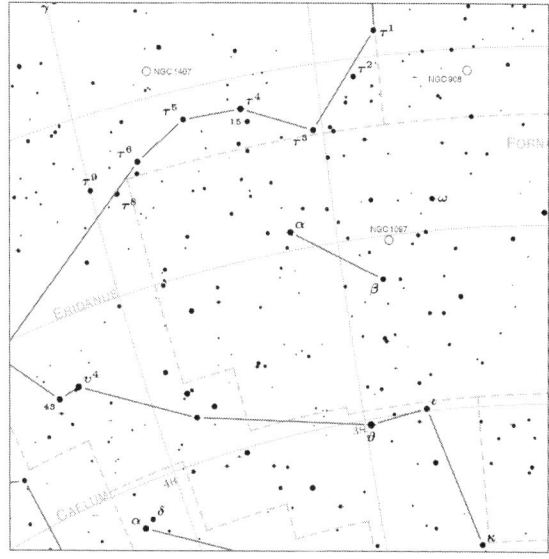

Eridnus

In Celtic lore the River Severn was recognised an important barrier which held back Saxon invasion for some time. Three great rivers of Wales actually are mentioned in some stories. One such tale tells of three sisters, who were spirits of the waters, who met at the mountain of Plynlimon to discuss the problem of what was the best way to the sea. The first sister took the direct route and headed westward to become the river Ystwyth. The second sister loved the landscape, and made her way through the hills and valleys to become the river Wye. The last of the sisters decided against short cuts meandered across Wales and the marches before reaching the sea. This Celtic spirit was known as Sabrina and her river is the Severn.

* * *

The area around β Eridani is dotted with faint galaxies, of which the Sc type NGC 1084, at RA 02h 46m 11s Dec -07°33m, is the brightest at magnitude

11.1. Close to β Eridani is the faint nebulae I.C. 2118, otherwise known as the 'Witch Head' nebulae, and possibly illuminated by Rigel, but it is beyond average equipment although it may prove fruitful as a photographic target.

Despite the fact that there is little of interest to the deep sky observer, there are several stars worth noting in the sprawl of this ancient constellation. o Eridani is a relatively close star to our own system and is a beautiful triple star. Additionally, this system contains one of the best white dwarf stars visible in a small telescope. Both companions of o Eridani are dwarf stars: one is the white dwarf mentioned above and the other is a lovely contrasting red dwarf. The magnitudes of these stars are 9.7 for the white dwarf and 10.8 for the red dwarf, whilst the primary star shines at magnitude 4.8, which makes it an easy naked eye object.

There is one beautiful planetary nebulae; NGC 1535 at RA: 04h 14m 12s Dec -12°44m. It is a slightly oval, filled ring of turquoise light of 9th magnitude and is instantly visible in the field as its position is quite far from the Milky Way. NGC 1535 takes a little effort to locate but the effort is well rewarded.

ε Eridani is one of the closest stars to Earth, and is remarkably close to the Sun in type, although it has about a third of the Sun's mass. It has been the target of enthusiastic professional astronomers searching for other life forms, as its Sun-like characteristics are held to be a good omen for alternative intelligent beings. This star underwent intensive 'listening' studies in the

NGC 1535

S.E.T.I. program during the late 1970s, but if anyone is out there, they have yet to signal their presence. The star is only 10.5 light years away and is visible near Orion's Belt. In 1998 it was found to have one Jupiter-like planet and the system is undergoing further scrutiny in order to discover any others.

Many of the faint stars in Eridanus are relatively nearby objects, some lying within a 25 light year range. The future of the human race may lie amongst the commonplace stars of this unusual constellation; they are close enough to be stepping-stones to the rest of our galaxy when our technology catches up with our dreams. On a planet circling ϵ Eridani, perhaps intelligent life forms not too different from us are similarly contemplating the beauty of our universe, and wondering 'is there anybody out there?'.

Gemini
(Gwyn, Gwythyr and the Lady of May)

The constellation of Gemini is seen by the Welsh not as twins but as two men battling over the love of a woman. They are Gwyn and Gwythyr, the sons of Greidawl who want the hand of the lady in red, Creudyladd. Ladies in red may now be perceived as of dubious repute, but to the ancients red was the colour of a bride, an outward expression of her virginity.

In the *Mabinogion* tale of Culhwch and Olwen, Gwythyr is mentioned as being part of the retinue of the legendary King Arthur. This myth has broadened to become synonymous with Sir Gawain and the Green Knight in Arthurian legend and is referred to in the folk song, '*Green grow the rushes oh*'.

In Welsh stories, a rivalry starts when Gwyn abducts his sister Creiddylad from her husband Gwythwr ap Greidawl. To get her back, Gwythyr amassed an army against Gwyn, leading to a battle between them. Gwyn was victorious and, following the conflict, captured a number of Gwythr's noblemen including Nwython and his son Cyledr. Gwyn would later murder Nwython, and by subterfuge force Cyledr to eat his father's heart in a barbaric act of cannibalism. After the intervention of King Arthur, Gwyn and Gwythr agreed to fight for Creiddylad every May Day until Judgement Day arrived. The one who was victorious on this final day would at last take the maiden. According to the story of Culhwch and Olwen, Gwyn was then placed over the brood of devils in Annwn lest they should destroy the race of men, and consequently Gwyn became the leader of the otherworld or the Tylwyth Teg. As such, Gwyn controlled the Cwn Annwn (see Canes Venatici) the baleful dogs whose howling foretells the death of anyone who hears them.

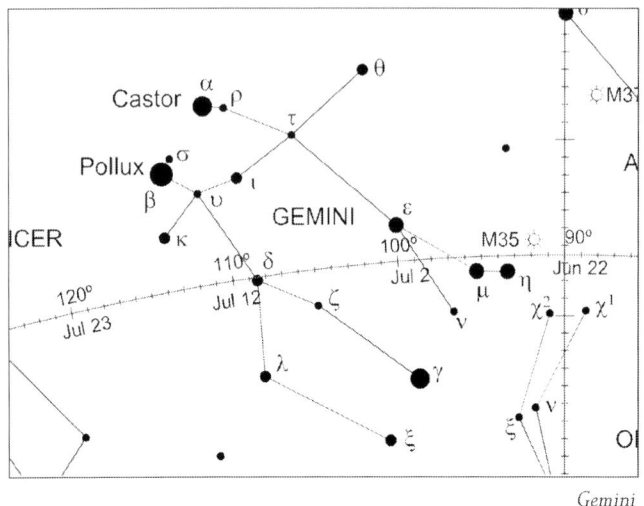

Gemini

In Celtic tradition, Gwyn and Gwyrthr are known as the Rivals of May and are interpreted as being the light and dark halves of the year battling each other as they set at midnight on May 1st, the Celtic feast of Beltane, the start of the summer season and hence important agriculturally for the growth of crops.

* * *

Gemini is unmistakeable; the twin rows of stars ending in opposite 'turns' at the feet of the constellation are obvious beacons on a winter's night. In addition, the two stars named Castor and Pollux are of similar brightness yet are of different colours. Castor is a brilliant white, whilst Pollux has a reddish hue. Despite the fact that Bayer assigned Castor the appellation α Geminorum, it is in fact Pollux that is the brighter of the pair, although this is not readily apparent without close scrutiny.

Castor is a binary star, one of the greatest binaries known, as no less than five companions have been recorded both optically and spectroscopically. Its most obvious companion is now quite close to its parent star, a very small separation of 1.9" makes it difficult to spot in a small telescope, but it has a magnitude of 2.8, making it surprisingly bright.

Gemini lies along the Milky Way, and is therefore packed with objects of interest to the amateur observer. Coloured stars, and doubles can be found in profusion throughout the group, and several star clusters are worth searching for.

The most famous of these clusters is M35, close to the northern foot of the twins. Two beautiful red stars lead onto this cluster if you wish to look through binoculars, but the sight of this region of star drops is unsurpassed through a telescope. M35 is a very rich cluster, roughly the size of the full moon. Star chains radiate out of the compressed centre, and flashes of light sparkle like diamonds against the backdrop of night as stars come into view. The cluster is situated at a distance of 2800 light years, and contains upwards of 150 stars or more, most of which are blue or white giants to be visible over such a vast distance with such brilliance.

Messier 35

M35 is one of the showpieces of the winter sky, a cluster that will delight over many years observation. Examining the surrounding field of M35, you will notice a faint patch of stardust to the southwest which is the galactic cluster NGC 2158. This luminous gauze of light lies at least twice the distance of M35, which is a great pity; if it were as close, then it would overshadow its famous neighbour. NGC 2158 seems to be an intermediate type of cluster, midway between a globular and a galactic cluster. It is extremely compressed and rich, containing over 500 suns in a very small area of space, shining with a combined magnitude of 9.2. It is not easy to resolve with a small telescope, but it should be easily visible nevertheless.

Not far from this group of clusters, along the foot of the northern twin, is an object that will not be visible to amateurs, but may well show up in a photographic exposure of the area. This is the nebulae I.C. 443, a crescent shaped gaseous mass of diffuse light. This object is the remains of a star that exploded as a supernova over 50,000 years ago. Astronomers are baffled and delighted by its shape, it is as if the explosion was all one sided. Inequalities in the interstellar medium may be responsible for this appearance. In long exposure photographs, it shows up as a lovely orange-red wisp of light.

One of the most enigmatic objects in Gemini is the planetary nebulae NGC 2392 at RA 07h 29m 25s Dec 20°54m. This small planetary has been nicknamed the Eskimo Nebulae, as in large telescopes the gaseous rings of matter surrounding the central star appear to look like a fur lined hood, whilst condensations of gas in the centre of the nebulae outline a crude face. The nebula is about 9th magnitude, but is easy to see lying about midway between $\varkappa$ and λ Geminorum. In a small telescope with low power it looks like an out of focus star, a light blue in colour. Once located, good seeing conditions and a high power may reveal some of this startling detail. It is a very distant object, apparently lying over 3000 light years away. Its high luminosity is doubtless due to the huge amount of ultra violet radiation causing the gas shell to fluoresce energetically.

One object well worth watching as a long-term project is the star U Geminorum. This is a very faint (magnitude 14) red dwarf star, mostly beyond the reach of average amateur equipment. However, it belongs to a class of objects known as Cataclysmic Variables, and can brighten by up to 5 magnitudes in a short space of time. The field to explore and find this elusive object is slightly to the north of 85 Geminorum. This star will be difficult to spot, if you can see it at all, as it lies in a rich portion of the Milky Way. Yet, if you learn the star field well enough, despite not seeing U Geminorum, when it does put in an appearance you will see it straight away.

Other objects of interest in Gemini include two small rich star clusters NGC 2266 and NGC 2420. Both clusters are faint and compressed, containing about 50 stars each. They ranges from 11 to 15th magnitude thus making them rather unremarkable objects for a small telescope. An extrasolar planet also resides in the constellation. The star HD 59686 has a planet six times the mass of Jupiter in orbit at a distance of and orbit of 303 days. The coordinates are RA 07h 31m 48s, Dec17 05m 09s and the star shines at magnitude 6.5 making this an easy binocular object. The stellar spectrum is K2III so the star should show as slightly orange.

For those observers using binoculars, the Milky Way in this region is well worth scanning. The presence of millions of stars will show up as a faint blue white mottled background against which the bright gems of the Winter Milky Way will be displayed prominently. Lose yourself as you reflect upon the treasures of this ancient constellation.

Lepus
(Melangell and Richard the tailor)

The constellation of Lepus – The Hare – is a sprawling smattering of stars directly to the south of Orion, and presumably is one of the hunter's kills. It is a motley tangle of stars that bear no resemblance to hare at all. Lepus is another constellation that was recognized by Ptolemy; the Babylonians also saw it as the figure of a hare and so did the Egyptians, who considered the hare to be an incarnation of one of their gods, and in their earliest astronomical tales it was associated with the boat of Osiris.

From the legends of west Wales comes a tale of a brave princess, Melangell who had left her home in Ireland and lived in a small cave in a hidden valley. One day, Melangell sheltered a hare from the huntsmen and dogs of Prince Brochwel. The hare was at its end but hid in the lady's skirts as she knelt in devotion saying her prayers. Impressed by her bravery in sheltering the animal, and her piety undisturbed by the snarling dogs, wheeling horses and the shouts of the huntsmen, Prince Brochwel gave the valley to Melangell to shelter the frightened and weak, be they man, woman or animal.

Over time, pilgrims made long journeys to rest at the tiny chapel she built in the valley and Melangell lived to old age as the abbess of a holy commune that is now associated with Pennant Melangell, or St Melangell's church near Llangynog in Powys. It is still a place of pilgrimage and the little church is a beautiful building in this mist-filled, remote valley.

In Breconshire legends this constellation represents the tale of Richard the Tailor of Llangattock. He was not only the tailor of the area but also the landlord of an inn there and was suspected of consorting with the Twlwyth Teg, and of practicing witchcraft. One day, a company of men were hunting in that vicinity, when the hounds startled a hare, which ran so long and so hard that they all tired of the chase and the hare disappeared from view into the cellar window of the inn kept by Richard the Tailor.

The hunters suspected that the hare was none other than Richard the Tailor himself, who could take different forms of animals as part of his association with the fairy world, and that his purpose in taking that form had been to lead them to the door of his inn and make them spend their money there.

* * *

Lepus is best viewed through the wide field of a pair of binoculars, as it is sufficiently small enough, and southerly enough to be explored in this fashion. Its stars are not brilliant objects, but they are visible without difficulty and can be used as stepping-stones to the two deep sky delights that await the observer.

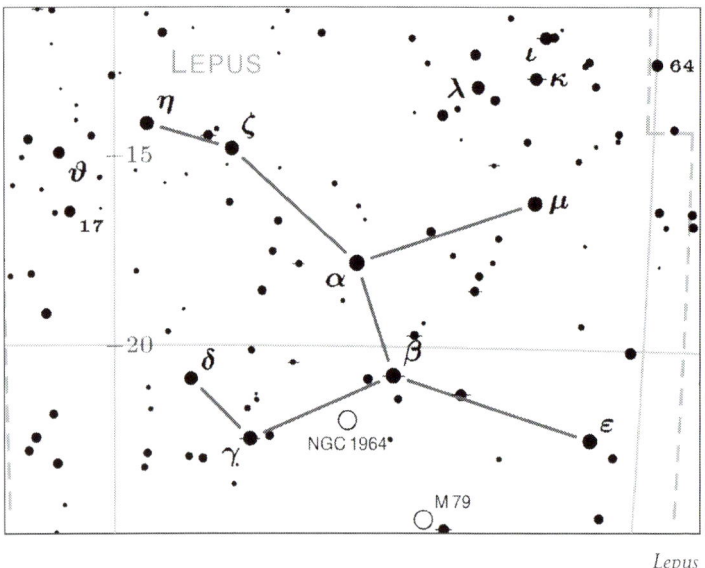

Lepus

One of the most attractive stars in this constellation can be seen with either binoculars or a small telescope. This is R. Leporis, a red giant of variable aspect, which changes from magnitude 6.7 to 9.8 in a period of 432 days. Once it has been found, its variability can be measured over a long period of time. Its allure lies in its deep red colour which at the time of the star's minimum light output, has been described as 'crimson', 'wine-red' and, to J.R. Hind the variable star observer, an "illuminated drop of blood". Its colour is certainly comparable to Herschel's 'Garnet Star' in Cepheus, although like all red giant variables, the colour fades slightly as the magnitude increases.

R. Leporis is one of a rare class of stars that are approaching the end of their lives. It is a carbon star, in that its spectrum reveals wide bands of absorbing carbon dust grains in its atmosphere, plus the presence of titanium oxide in molecular form in its outer layers. The surface of the star therefore has a low temperature of about 2500 degrees Kelvin. The distance to R. Leporis is not known with accuracy, but the best interpretation of the current data puts the star at 1300 light years.

The other notable attraction of this constellation is the globular cluster M79, which was discovered by Messier in 1789. It is not a particularly impressive sight from our latitudes, but it can be seen with a pair of binoculars, whilst a small telescope will reveal a subtle mottling of half resolved stars. M79 appears to become more compressed towards its centre, and shines as an 8.5 magnitude studding of starry points. The cluster seems to be about 41,000 light years distant and possibly contains upwards of 90,000 suns. M79 can be found by drawing an imaginary line through α and β Leporis and extending about 4.5 degrees to the south. The cluster lies close to an interesting fifth magnitude double star H 3752.

Lepus contains a rather poor star cluster, NGC 2017 centred on a multiple star just to the south east of α. It is a large scattered cluster of 20 unremarkable stars, best seen in a telescope.

Lynx
(The Otherworld Guardian)

The cat in Celtic mythology was often associated with the Otherworld and was an animal sacred to the moon goddess. Red, white and black cats were of special interest to the Celts and tales of crossing the path of a black cat and the ill-luck that comes with it still occur today. The ancient tribe of the Silures of south Wales venerated the wildcat and it is known that the lynx was a favourite of the tribe as in early Welsh history this cat could be found all over Europe.

Lynx represents the guardians of the Otherworld; silent and mysterious they are well suited to this role. They keep the secrets of the otherworld to themselves, as they gaze with guile upon a world that does not comprehend the depth of their knowledge. When invoked, they can grant the caller a variety of insights regarding more esoteric, ethereal knowledge.

The lynx, has a heavy body, tufted ears, heavy side whiskers and green-gold eyes. They are night hunters and prefer to spend their days resting in a cool cave or other lair. The lynx was an alternative form of the god Llew Llaw

Gyffes and the animal is not far from the hero in the sky and sharing a small part of the Milky Way in its southern reaches.

* * *

Lynx is a rather obscure collection of stars, most of which are 5th magnitude objects, and widely spaced in the sky. The constellation lies to the north east of Gemini, bordering on Auriga, but contains few objects of note. The group is possibly an ancient constellation as it is mentioned by Aratus in the second century BC, being positioned in front of the great bear. Its modern pattern was created by Hevelius to fill a convenient gap between the constellations of the winter sky and those of spring.

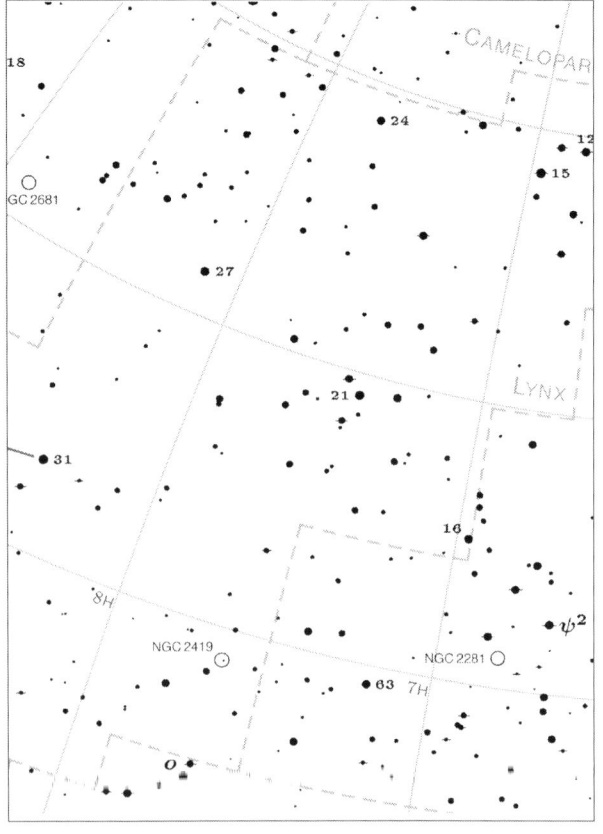

Lynx

Lynx contains two objects of note, although the observer equipped with binoculars had better move on, or simply note the position of this constellation, as these objects are beyond such modest instruments. The first deep sky wonder is a lovely edge on spiral galaxy of Sb type, NGC 2683, which lies just above the borderline with cancer, west of the star Alpha Lynxis and on a line with it. NGC 2683 is a 10.6 magnitude smudge of blue white light that is easy to spot in a low power eyepiece, the most difficult part is finding it among the barren stars of this group!

The most observed object in this constellation is a remote globular star cluster NGC 2419, which can be located close to the 5th magnitude star ψ Lynxis, above the glowing luminance of Castor in Gemini. This is not the greatest cluster, in fact it is one of the dimmest, shining as it does at magnitude 11.2. Its stars are virtually irresolvable in a small telescope and it is in fact the most distant globular cluster known to belong to our Milky Way galaxy.

NGC 2419 was discovered by William Herschel in 1788, but its actual distance was not known until earlier this century when the astronomer Harlow Shapley calculated that it lies an enormous 278,000 light years from us, over 300,000 light years from the focus of its orbit, our galactic centre. NGC 2419 contains over 175,000 Sun-like stars, making it a respectable cluster despite its remoteness.

About 150 globular clusters are known to orbit our galaxy, but they mostly lie within a sphere of 65,000 light years of the galactic nucleus. Why NGC 2419 should be different is a mystery. The theory has been put forward that it does not really belong to our galaxy at all but is an 'intergalactic tramp' or wanderer through the galaxies of the local group. However, studies of its recessional speed indicate that it may not have sufficient velocity to escape the gravitational tug of the Milky Way, and in a few million years it may return to us, charging through space at quite a rate as it obeys Kepler's third law of orbital motion.

NGC 2419 is not the only globular cluster to lie at such a remote distance. The other object known to be at a comparable distance is NGC 7006, a globular cluster in Delphinus, lying in the opposite direction to this one in Lynx. That being the case, the separation of these clusters must be huge, at least 5 times the diameter of the disc of the Milky Way. Make a diligent search therefore for both these remote objects, revelling in their independence from our galactic home.

Monoceros (The Unicorn of Peredur)

Monoceros – The Unicorn – is an ambiguous constellation that is very difficult to discern as it has few bright stars. It's claim as a constellation of antiquity is debatable though it does appear on some old Arabic maps and spheres from the twelfth century and therefore may possibly be recognised by earlier cultures.

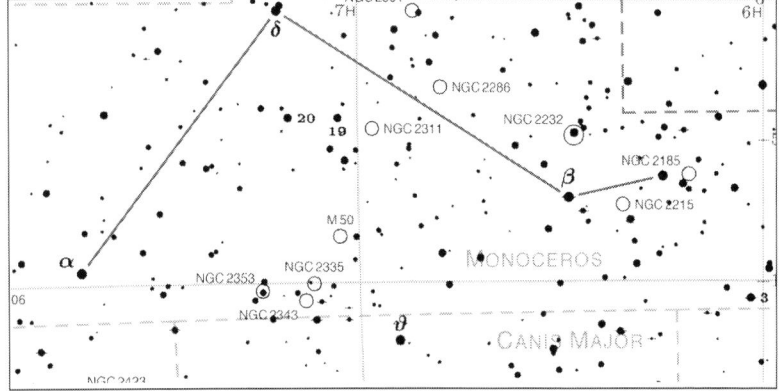

Monoceros

In the legends of the dark land of Wales it appears to be involved with the Arthurian knight Peredur, or Sir Percival, who is recorded as the son of Efrawg. It is whilst he is on one of the Arthurian quests that Peredur stays with the Witches of Gloucester and there has to undergo a series of tests, or tasks, almost like the mythical Hercules and his twelve labours. In one of the tasks Peredur encounters a unicorn, which charges him but which he sidesteps and kills the animal, cutting it in two. He is chided for this by one of the witches and her curse is undone only when he duels with the infamous black knight.

In this tale from the *Mabinogion*, the unicorn is portrayed as a very strange animal which appears unrelated to any other, suggesting that it is otherworldly and perhaps in concert with the witches themselves. It is a heavenly representation of something that is out of this world, perhaps a creature from the spirit realm that can bridge the gap between earth and heaven. Jonathan Miles-Watson in his book *Welsh Mythology, A Neostructuralist Analysis* claims that the unicorn represents a monolithic but mysterious structure in which the two halves of the unicorn – horse and mythical creature perhaps represent male and female halves making a whole.

The current constellation was created by Hevelius to fill in a blank space between Orion to the west and Cancer to the east. Thankfully, although its stars are few, its deep sky objects are not, as Monoceros lies in a particularly fruitful part of the Milky Way.

This richness is derived from a huge molecular cloud of gas and dust that permeates this region of space and has outcroppings in the form of bright gaseous nebulae that make Monoceros a treasure house for the well-equipped observer. Do not despair if you own binoculars or a small telescope, as there are plenty of objects to tantalize and astonish.

One of the best star clusters of the winter sky is to be found in the southern part of Monoceros. M50 is a bright nebulous patch of light in the field of a pair of binoculars, but under the scrutiny of a small telescope, it becomes a treasure trove of over 200 stars in a small compressed area of space. Not all these stars will be visible, but the primary stars are easy objects to see. One of them is a delightful deep red in colour and is immediately recognizable even in a small scope. The distance to this lovely cluster is 3000 light years, thus making the stars that create its 6th magnitude glow very luminous indeed. Its position in a lavish part of the Milky Way assures M50 of a unique place in the memory of those that observe it.

Messier 50

A most beautiful cluster plus an attendant nebula is the next stopping point on our tour of Monoceros. In a tight group around the star 15 Monocerotis, is a wonderful pack of glittering points of light, all caught in a misty web of faint light. This is the cluster NGC 2244 and nebulae NGC 2237, otherwise known as the Rosette Nebulae. It is a fantastic sight in giant binoculars on a clear night, but is a disappointment to those with small telescopes. The cluster of 40 or so stars is readily apparent, but in the confined field of an eyepiece, the nebula lacks structure and disappears altogether. The Rosette Nebulae is easily captured on photographic film, and is a beautiful orange red in colour, which contrasts wonderfully against the electric blue starlight of the cluster.

The Rosette Nebulae lies at an approximate distance of 5200 light years, giving the cloud of gas a dimension of 55 light years in extent. The nebulae has been said to contain enough matter to form 11,000 Suns, and indeed such stellar births are still occurring in this magical region of our galaxy. Large telescopes show several knotty condensations of dark matter contrasting with the ruddy hue of the gas. This is where the next generation of stars is currently being formed, and the area is under intense scrutiny by astronomers.

Further to the north of this object is a lovely cluster around the star S Monocerotis. This cluster, NGC 2264 is commonly known as the 'Christmas Tree' cluster, for reasons that become obvious the moment one views it. About 25 bright stars make up the illuminated tree, most of which can be seen under good seeing conditions through binoculars, although a telescope gives a finer view. This group lies at a similar distance to the Rosette Nebulae adding further proof that the region is alive with stellar nurseries. NGC 2264 is also surrounded by faint nebulae, but this nebula is reserved for those owners of large telescopes, as its surface brightness is very low.

A little to the south of this cluster lies the enigmatic object known as R Monocerotis. This is a variable star of very unusual type as it appears that it is struggling to throw off its swaddling bands of gas and dust and emerge as a young main sequence object. However, it seems to be experiencing difficulty as, occasionally, R Monocerotis disappears from view behind a rather interesting faint nebula first discovered by the eminent astronomer Edwin Hubble. This object is easily visible in a small telescope to the southwest of the Christmas Tree cluster as a small smudge of white light amongst a rich field of stars.

The only fairly bright star of the constellation, β Monocerotis is also a fine triple star, one of the most splendid stars in the winter sky. The components are widely separated and of almost equal magnitude range, all around mag. 5.5, and of similar colours.

Being a part of the Milky Way, Monoceros abounds in star clusters, many of which are easily visible in binoculars or a small telescope. The only problem the observer has is trying to find them among the plethora of faint stars in this extremely rich region. One of the nicest of these clusters is NGC 2301 at RA 06h 51m 48s Dec 00°28m, a fantastic arrangement of over 60 stars in a compressed group. Most are relatively bright at mag 8, so the cluster should be visible as a misty patch in a good pair of binoculars.

A slightly fainter but nevertheless rich object is NGC 2324, which contains 50 stars of around 10th magnitude in a small, condensed group that is a pleasant sight. One of the best-known star clusters in Monoceros is NGC 2506, at coordinates RA 08h 00m 12s Dec -10°47m; a beautiful gathering of 75 stars in a large group. The stars look like tiny needle points of light, shining at around mag 11, but the overall magnitude of the cluster is around mag 9, so it isn't difficult to spot.

Scanning the Monoceros Milky Way with binoculars is rewarding. Mounted on a tripod, these instruments make effective wide field telescopes that can be comfortable to use. The constellation of Monoceros is filled with the interplay of light and darkness, which binoculars reveal, as dust clouds and star clouds compete for the observer's attention.

Orion
(Mabon and Owain Llaw Goch)

The ancient Celts knew the constellation of Orion as Cernunnos who was represented as sitting crossed legged, with antlers on his head and a torc of gold around his neck. The role of Orion as Cernunnos in English folk tales is played by Herne the Hunter of Windsor Forest but in Welsh mythology he is Mabon, the deity of the winter sun, who is the only god who can handle the hunting dog Drudwyn. Mabon was rescued by some of the Arthurian knights in the tale of Culhwch and Olwen and assists Culhwch to hunt down the Twrch Trwyth (woodland boar or the constellation of Leo) so that he can win the hand of Olwen from the giant Ysbaddaden. It would seem that the Welsh association with Arthurian legends is well set as Mabon is heralded as one of the retainers and servants of Uther Pendragon and is mentioned by Taliesin.

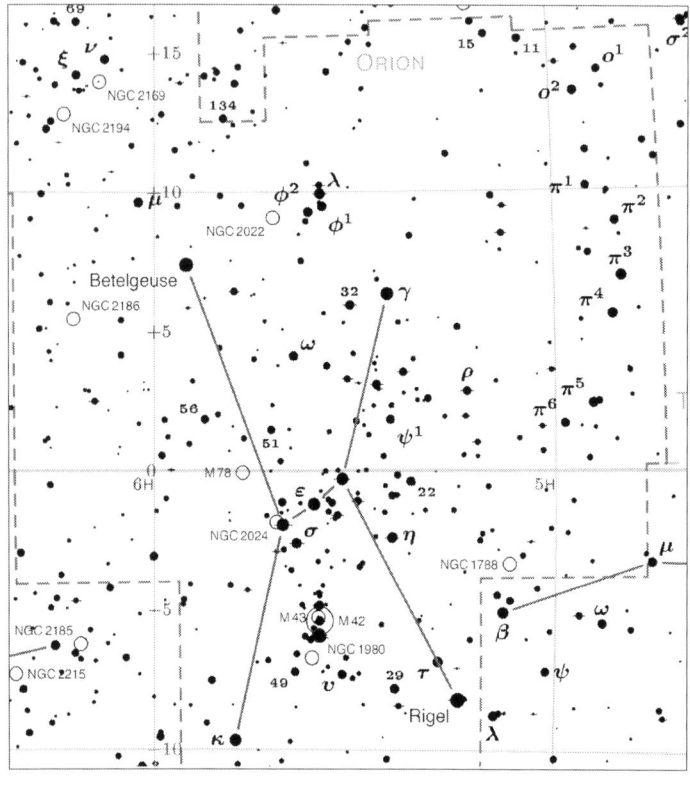

Orion

Orion is also associated with the legend of Craig y Ddinas or Dinas Rock, a popular climbing spot in the waterfall country on the edge of the Brecon Beacons National Park. The tale has it that a drover called Dafydd Meurig took his cattle to London and encountered a stranger who asked him to show where his hazel staff had been cut. The stranger accompanied Dafydd to a tree near Dinas Rock. When Dafydd pointed out the tree, the old stranger moved to the side of it where he lifted a large stone hidden in its roots and both men went down into the earth. There they found a great giant, Owain Llaw Goch, surrounded by treasure and armoured men. The stranger informed Dafydd that when the giant awoke, he would be the King of Britain but would sleep for now, being watched over by his giant counterpart in the sky, the group we know of as Orion. Dafydd became very wealthy as he kept part of the treasure but would never reveal the source of his wealth under Dinas Rock.

* * *

The seven stars that make up the prosaic pattern we see today are relatively young, perhaps only a few million years old. α Orionis is a possible exception to this rule, a red giant nearing the end of its life. In addition, it appears to lie much closer than the other stars of the group, thus suggesting its evolutionary independence. At a distance of 900 light years, it contrasts with the 1500 light years for the rest of the group. The name given to this beautiful star is Betelgeuse, meaning 'armpit of the giant', due to its situation under the upraised arm of the hunter. Betelgeuse has a measured diameter of over 1 billion km and has a tenuous atmosphere of potassium gas larger than this again, making it one of the largest stars known. Additionally, Betelgeuse is slightly variable, thus it can be marginally brighter or dimmer than its opposite companion Rigel.

Rigel or β Orionis, is an amazing star, a veritable searchlight among its stellar compatriots as it shines with a luminosity of around 60,000 times that of the Sun. It is classified as a blue supergiant of type B8Ia and has a mass of 30 times that of our Sun. Rigel is considered the principal illuminator of a nearby reflection nebula, the 'Witch Head' nebulae, but this gaseous cloud is usually too faint to pick up with small amateur equipment.

The two rather unremarkable stars that constitute the remaining outline of the hunter are Bellatrix, marking the western shoulder, and Saiph, in the position of the eastern knee. Both are blue giant stars, but the eye is captured by what lies between them, the three stars that make up the unforgettable belt of the hunter.

These stars in the ascending order from east to west are Alnilam, Alnitak and Mintaka. They can be perceived together in a line, and make a lovely target for binoculars. In ancient Wales the stars were known variously as Llathen Teiliwr, the tailors yard, perhaps in reference to the straight line, like a rule that would measure a bolt of cloth or Y Tri Brenin: 'the three kings', or as Y Groes Fendigiad, 'the blessed cross', both the latter names probably refer to the story of Jesus in early Welsh Christianity.

Lying to the south of this belt of stars is one of the most famous objects in the heavens, the great Orion Nebulae, otherwise known by its designation Messier 42.

The Orion nebula is an enormous cloud of gas and dust over 26 light years across lying on the nearside of the Orion arm of our Milky Way galaxy, around 1350 light years away. It is easily visible to the naked eye, and is a magnificent sight in a pair of binoculars, which reveal it to be a glowing patch of misty light surrounding a small cluster of bright stars. The view through a telescope is even more remarkable; a great ghostly blue white glow pervades the eyepiece, looking like a giant bat gliding out of the darkness. In the centre of this illuminated mass is one of the most beautiful multiple stars in the heavens, θ Orionis, known as the Trapezium. This quadrangle of stars is visible even in a small telescope, and as one looks closer, other stars flash into view, giving the impression of a spangled field of light. The area of M42 contains over 150 newborn stars, many of which are targets for variable star observers.

Messier 43

On the northern edge of the nebulae is a dark bay commonly called the 'Fish Mouth'. This a dark intrusion of dust arises from the huge molecular cloud of which the Orion nebulae is the brightest portion. The fish mouth effectively separates M42 from its small companion M43, although to most observers M43 is an extension of M42. Many a eulogy has been written of this nebula, so explore it. Interestingly, for those interested in extrasolar planets, the nebula contains a number of 'rogue' planets not affiliated to one star. They may have been thrown out of their systems by interactions, but little more is known about them apart from their massive size.

Just to the north of the M42/M43 group is a wispy reflection nebulae surrounding a small star cluster designated NGC 1977. The stars are easier to find than the faint nebulae but the nebulae may show up well on a long exposure photograph. Heading to the north of the Orion's Belt, there is a faint nebulae, M78 to the north of Alnitak. M78 is not bright, although it can be seen with a small telescope quite easily, a ninth magnitude star which illuminates one edge of this little fan-shaped nebulae and which looks slightly green in colour. The whole of the constellation is shrouded in wreaths and streamers of gas, most of which are too faint to be seen with the eye, even using a telescope.

One particular object worth studying is the 'Horse Head Nebula' or Barnard 33. This is a dark patch of dust obscuring a bright nebula behind it, looking just like the knight chess piece. The nebulae lie just below Alnitak in the belt but are not usually visible in amateur equipment. However, it shows up well in photographs, as does its companion, the Flaming Heart Nebulae, or NGC 2024 to the east. Further afield is a large wisp of gaseous matter known as Barnard's Loop. This is impossible to see with average equipment, although, once again, a short exposure will enable you to capture this elusive object.

Orion is constantly undergoing change, albeit slowly. Several observers have remarked upon nebulae close to the star $\pi 6$, part of the western arm of Orion, depicted holding the lion's skin as a shield. This has been called the Peekaboo Nebulae as it has brightened and faded over recent years. It is thought that this nebula marks the position of a star emerging from its stellar nursery, throwing off its surrounding dust.

Puppis
(The Ship Prydwen)

This constellation does not appear in its entirety when viewed from Britain, but some of its stars are to be found to the east of Canis Major, where ρ Puppis is the brightest star. Perhaps some of it was known as a whole ship, the *Argo Navis,* by early Welsh bards and so it become part of legend as *Prydwen,* the ship of King Arthur, according to the Welsh poem, the Spoils of Annwfn by Taliesin. This ship also appears in Culhwch and Olwen tale, and is used by Arthur to travel to Ireland in order to obtain the cauldron of Diwrnach and the boar Twrch Trwyth.

Ships of all shapes and sizes from coracles to sea-going barks were known to the ancient Welsh and even into the seventeenth century it was quicker and safer to travel around Wales (and go to Cornwall, Ireland or Scotland) by ship than travel overland. However, the Welsh are not known for daring sea-going adventures – except for one.

Prince Madoc was an heir to the kingdom of Gwynedd and after the death of the king Owain Gwynedd, Madoc was dismayed by the family in-fighting for the throne. Madoc and his brother Rhiryd took their families and those who attended them out to sea and sailed westward from Llandrillo, or modern

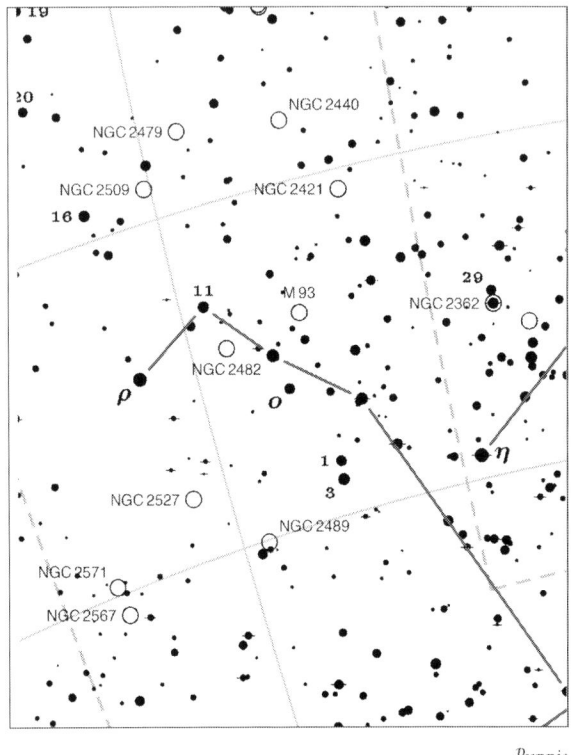

Puppis

day Rhos-on-Sea. In the year 1170 they apparently discovered a distant land and settled there permanently. They later sent word back about the abundance of food there and tempted other settlers from Wales to join them.

The story is in all probability a medieval fable but it was used by both Welsh and English to claim dominance over the New World and halt Spanish claims to the territory in the fifteenth century. Throughout north America, there are towns named Madoc in his memory and at Fort Morgan in Mobile, Alabama, there is a plaque extolling the landing of Prince Madoc.

* * *

Puppis is now a segregated part of a much larger constellation named Argo Navis in Ptolemy's time. This is the legendary ship which took Jason on his quest for the Golden Fleece, and Puppis represents the poop, or quarterdeck, on which Jason would have stood to guide the ship. The rest of the ship, consisting of the constellations Carina and Vela are invisible from Britain. Puppis is noteworthy due to its prominent position along the Milky Way, and as one would expect, the area is littered with bright star clusters.

Unfortunately, only three of these clusters are well seen from our latitudes, but they are among the most incredible of objects to be seen in the sky. The best cluster is undoubtedly M46, a large grouping of over 500 stars, of which at least 100 are visible in a good pair of binoculars or a small telescope. The field is quite scattered, not compressing towards the centre, but radiating star chains in all directions. On close inspection the northern edge of the cluster reveals an out of focus star. This is the wonderful planetary nebulae NGC 2438, which shines as an 8th magnitude patch of fuzz. The distance to M46 is around 5500 light years, but the nebula is closer to us, making this a lovely accident of nature. Even so, the planetary nebula is very distant, lying around 3300 light years away. This makes it a large planetary with a diameter of just over one light year, containing a faint central star of 16th magnitude.

Just to the west of M46 is another remarkable Messier object, the cluster M47, an aggregation of 30 bright stars, easily visible in a pair of binoculars, and a visual treat in a small telescope, as most of the central stars shine brightly

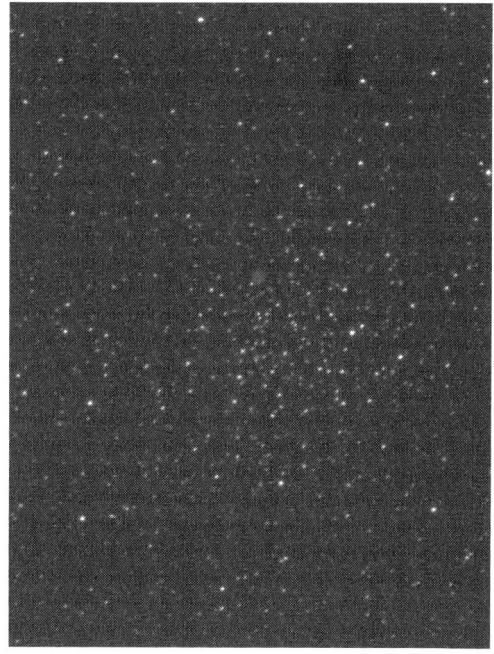

Messier 46 with NGC 2438

at magnitude 6. M47 is closer to us than its spectacular neighbour M46 a computed distance of 1600 light years. At the heart of the cluster is a wide double star.

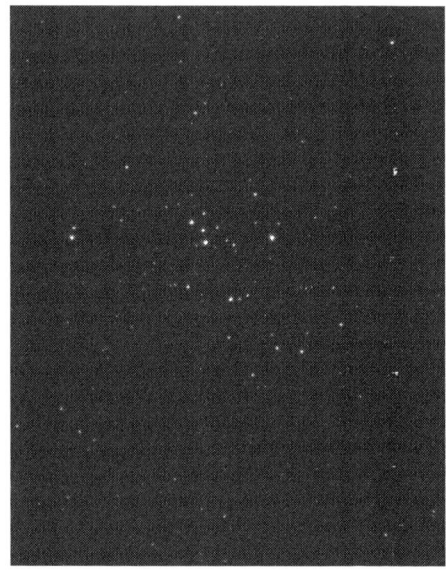

Messier 47

Another object of note is the star cluster NGC 2539, which lies around the double star 19 Puppis. It is a rich compact assembly of over 100 stars that shine at magnitude 11, giving the cluster an overall magnitude of around 9. The group can be seen with binoculars as a misty patch of light, but a small telescope will reveal more detail.

One other object worth striving for is the compact group M93, which lies 9 degrees south of M46. Binoculars will uncover it easily, but its southerly aspect drowns out what would be a wonderful sight. A telescope will, once again, improve the view of this cluster, which would be a naked eye object if it lay at a higher declination. M93 is a collection of over 60 suns at a distance of 3600 light years.

Although Puppis appears as a rather unremarkable constellation to northern sky observers, it is worth sweeping with binoculars, as small clumps and starry streams make this group a rich hunting ground for those with patience and dedication.

Taurus
(The Ox of Hu Gadarn)

The constellations of Taurus and Auriga are fused together into a large pattern in the sky which tells the tale of Hu Gadarn, the first man to link oxen to the plough. The entire assembly starts with Bootes the herdsman, the role taken in Welsh legend Hu Gadarn, moves in a line through Ursa Major and comes down through Auriga to Taurus the oxen. If Hu is handling the plough, then it is easy to see why the seven stars of Ursa Major are so named in Welsh tradition. On a dark winter's night, Taurus dwells high in the sky as seen from Wales, and contains several objects of note.

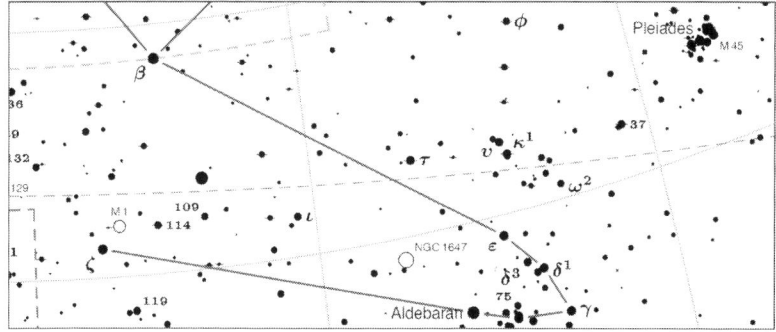

Taurus

The Celtic zodiac at one time began in Taurus which initiated the waxing half of the year. Belenos was the sun god of that part of the year. In the isle of Ynys Mon (Anglesey) ancient Druids sacrificed white bulls to the sun god and performed the bull feast as an important divining tool. The constellation is sometimes called the Ox, and in this regard was a castrated bull and a docile animal of great strength used to haul wagons or a plough.

The stellar grouping that forms the constellation of Taurus is one of the most easily recognizable associations in the sky. The wide V shape of the head of the bull, and its prominent star Aldebaran plus the beautiful cluster of the Pleiades makes Taurus one of the marvels of the night sky. The constellation dates from antiquity, and has been recognized as a heavenly bull by practically every civilization on the face of the Earth; indeed, it is one of few constellations that actually looks like the beast it represents. In Greek mythology the constellation is the bull that Zeus became to run off with the nymph Europa. Only the head and shoulders of the bull are depicted in the sky, and in this tale the bull bore Europa across a river, making the body invisible.

By far the greatest deep sky object is the Pleiades, a jewel like cluster of seven stars that glitter like diamonds. The cluster actually contains over 200 stars, but only the brightest ones are visible to the unaided eye. It represents the seven daughters of Atlas, the giant who held up the heavens from the Earth in Greek mythology. The cluster itself is mentioned in the literature of almost every culture and the Chinese made the first recorded observation in 2357 BCE, although it is possible that Babylonian scripts could go back further. The cluster is even mentioned in the King James Bible in the book of Job, chapter 38 vs 31.

The central star of the cluster, Alcyone, is the brightest object, ruling impassively over her blue supergiant sisters. The cluster is commonly believed to be only a few million years old, recently formed on the astronomical timescale, and this is borne out by the presence of shrouds of gas and dust that surround the stars with a blue nebulous glow. The poet Tennyson referred to the ethereal beauty of this group in *Locksley Hall*, quoting:

> Many a night from yonder ivied casement
> ere I went to rest,
> did I look on great Orion sloping
> slowly to the west.
> Many a night I watched the Pleiad's rising
> through the mellow shade,
> Glitter like a swarm of fireflies
> Tangled in a silver braid.

To those fortunate enough to own a telescope large enough to show this nebula visually, the description above fits the asterism exactly. The Pleiades lie about 420 light years away and are best seen with either a low power ocular, or even better, a pair of binoculars, which capture this fantastic group in one field. Close to Alcyone is a marvellous triple star that is readily apparent in a small telescope, whilst scattered around the field is an amazing assortment of bright stars and doubles. The Pleiades are a truly arresting sight, certainly one of nature's visual successes.

In Wales the stars are known as Saith Seren or Y Twr Tewdws, the seven stars or the 'thick group', and there is some reference to the stars as being Y Saith Seren Siriol, meaning the seven cheerful stars, though why they are cheerful is a minor mystery unless one contends that to the ancients the setting of the Pleiades with the sunset heralds the onset of summer and warmer days and promise of a good harvest to come.

Messier 45, The Pleiades

Not to be overlooked is the other major cluster of the constellation, the Hyades. This is the V shaped group of stars that make up the familiar outline of Taurus, and is remarkable in that it is the second closest star cluster to Earth, and hence is a stellar laboratory for cluster theories. The Hyades are approximately 130 light years away, and like the Pleiades, are best seen with binoculars; which reveal a wonderful group of stars ranging from 5th to 10th magnitude. Theta Tauri is a naked eye double; both components are a glorious yellow in colour, whilst most of the other stars of the cluster are white. The primary star, Aldebaran, is not a component of the Hyades as it lies 55 light years away, with the cluster making a pretty backdrop for this K type giant star, 45 times the diameter of our Sun.

The object that initiated Messier's catalogue is to be found just above the star marking the southern horn of the bull. The Crab Nebulae, so called by Lord Rosse, was first seen by Messier in 1758, but had been seen by other observers prior to that time. Little did they suspect the impact this object was to have on the future of astronomy, as it was the first confirmed remnant of a supernova, plus the first visually detected Pulsar in our universe. The Crab Nebula is an eighth magnitude smudge of blue white light in a small telescope, with a distinct S shape. Binoculars will be rather hard put to pick it up as it has a low surface brightness for an 8th magnitude object. Observers with larger telescopes may well be able to discern the filaments around the periphery of the nebulae, plus the few stars that appear to be embedded in the gas, although these are only field stars.

The Crab is the remnant of a star that was seen to explode in the year 1054 AD and has a well-documented history in the annals of both the Chinese and the Native Americans, although no sighting was made or recorded in Europe, presumably due to prevailing superstitions of the time, or perhaps the loss of records. It was an incredibly bright object, visible in the daytime sky for over three weeks, and visible at night for over a year. Today, astronomers ponder over this marvel of nature that has taught us so much about the universe we live in.

In the eastern portion of the constellation, close to the border with Auriga and lying amongst the stars of the Milky Way is an enigmatic object that is invisible to amateurs, but may be glimpsed with a long exposure photograph. This is another supernova remnant, termed S.147, and is a wreath like vapour of gas that is difficult to separate from the background Milky Way. It appears to have originated over 50,000 years ago, but no pulsar has yet been detected in this part of the sky, only a faint source of radio emission. Perhaps age has caused the pulsar to become inactive or fade away.

There are several other objects to scan within Taurus, notably two star clusters, NGC 1647 and NGC 1746, which both lie between the horns of the bull. NGC 1647 lies just a short way in a direct line from the lower arm of the V of the Hyades and is a 7th magnitude group of some 25 stars in a compressed field which can be glimpsed with binoculars. NGC 1746 lies nearer the ends of the horns of the bull and is a more scattered group containing about 50 stars of 8th magnitude upwards. Both groups are visible in small telescopes and are a pleasing sight. Another two star clusters of note can be found together in the same low power field, and are NGC 1807 and NGC 1817 respectively. The richer of the two is NGC 1817, which contains over 50 faint stars, whilst NGC 1807 is a small unremarkable group of 15 stars in a compact group.

CHAPTER FOUR
The Spring Constellations

'Is there not
A tongue in every star that talks with man?
And wooes him to be wise? Nor wooes in vain;
This dead of midnight is the noon of thought,
And wisdom mounts her zenith with the stars.

—Anna Laetitia Barbauld

Chapter Four –
The Spring Constellations

With winter disappearing fast, the astronomer looks forward to the skies of spring, which bring many delights to tantalise and enthral the observer. The only drawback with the skies of spring is that as the nights get lighter there is less time to observe the constellations and their attendant deep sky objects.

Spring is heralded by the rising of the constellation of Boötes (pronounced bow-oot-ays) and its brilliant first magnitude star Arcturus, a lovely orange giant lying fairly close to us in space. Arcturus is the second brightest star visible from this country, and dominates the sky from spring to late summer. The other constellations clustered fairly close to this assembly of stars are Corona Borealis, Hercules, Serpens and Ophiuchus, whilst the southwestern aspect of the spring sky is taken with Cancer, Leo, Hydra and Sextans. To the southeast are the constellations of Virgo and Coma Berenices with Corvus and Crater lying well to the south.

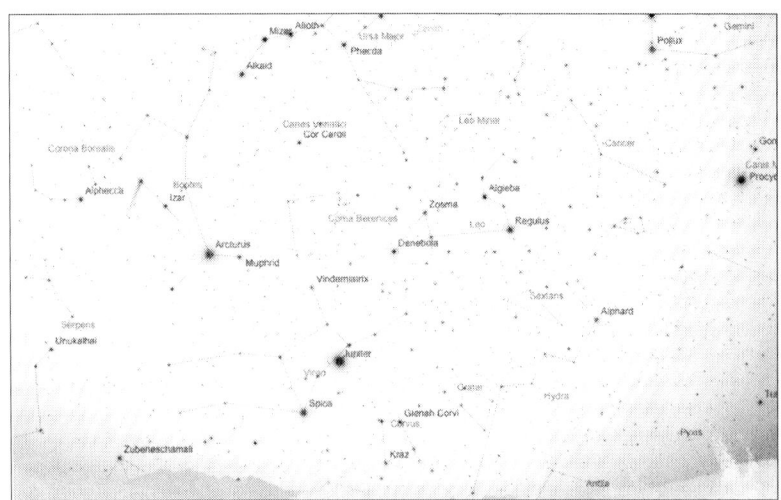

Spring constellations

The constellations of the spring sky cluster around the most familiar figure of Leo the lion. Its recognizable outline really does resemble the mythical Nemean lion and its stars are bright enough to render the constellation as the most conspicuous group. Below the lion is the straggling constellation of Hydra that winds from the winter to the summer sky and lying below and to the east of the lion is the Y-shaped constellation of Virgo. Above Virgo is a smattering of faint stardust that is the group of Coma Berenices whilst above and east of this group are Boötes, the arc of Corona Borealis and the fainter six stars of the constellation of Hercules. Lying directly above all of these, at the zenith, is the familiar constellation of Ursa Major leading into the circumpolar constellations.

During spring, the observer is looking out into deep space. The Milky Way forms a disc around the horizon and cannot be seen well until early morning. However, what is lost in the way of our home galaxy is certainly made up by the galaxies of the spring sky, which abound in such numbers as to be virtually uncountable. The constellations of Virgo and Coma Berenices are the area which the great American astronomer Edwin Hubble immortalised as the 'Realm of the Nebulae'. Hundreds of faint galaxies of the fabled Virgo Cluster can be seen on any one night by observers with patience, whilst one of the nearest star clusters makes up the shimmering constellation of Coma Berenices, an added bonus.

Just because the nights are getting lighter does not necessarily mean that they are warmer, so take the same precautions against the weather as you would in wintertime in order to enjoy the wonders of the spring sky. They are best viewed during the month of April as the sky becomes sufficiently dark to observe for several hours at a stretch. During May however, the sky does not really get dark at all, so you lose the advantage of contrast if you are looking for faint objects such as the wealth of galaxies in the spring groups. Better to stay up all night in April than have a frustrating few hours in May!

The Sun in late spring never dips more than 18 degrees below the horizon, so prepare well for your spring observing, and catch these objects at their best, against the dark sky of early spring. The end of May and the month of June find most of the sky washed out in temperate latitudes. As twilight arrives, look out for the bright stars of the spring groups as they appear in the gloaming.

Low on the horizon in the south can be glimpsed a few of the faint stars of two spring constellations; Centaurus the centaur and Lupus the wolf. From British latitudes there is nothing to note in either of these groups and none of their major stars rise northward of 40° latitude.

Distinctive Stars

Arcturus (spectral type K2III) is the brightest star north of the celestial equator, the brightest orange star, and the first-magnitude star with the most unusual motion and destination. Arcturus is the only bright star we see that is not travelling in the equatorial plane of the Milky Way galaxy, but instead is circling the galactic centre in an orbit highly inclined to that plane. Arcturus is presently swooping down through the equatorial plane, having been too far north of the plane to be visible to the naked eye from Earth about half a million years ago. Its future path brings us a little closer to it in the next few thousand years, but then its dive south of the equatorial plane takes it away from unaided vision again in about half a million years. The space velocity (true velocity through space) of this wandering star is twice as great as any other first-magnitude star, and about 3 to 10 times greater than most. Little of this swift motion is in radial velocity (the component of its motion directed away from or toward us). Thus, although it is about 36 light years away, its proper motion is much greater than that of any first-magnitude star except our nearest neighbour Alpha Centauri. Arcturus is also the star that marks the onset of autumn by its evening setting.

Arcturus is heading at the rate of 2.28 arc-seconds per year toward the territory of the constellation Virgo. Arcturus appears to have no companion, but its colour, variously described as 'topaz', 'golden-orange', and 'champagne-coloured', deserves careful study, and illustrates that colour is in the eye of the beholder. This wanderer is a great object, up to 15 times larger and 100 times more luminous than our Sun.

Spica (spectral type B1V) in Virgo has been shown to be a spectroscopic double with the companion star very close, slightly eclipsing the main star – and the main star itself subject to regular pulsations. The brightness changes associated with these slight eclipses and small pulsations are too minute to notice visually. Look for blue in the glow of Spica.

Regulus (spectral type B7V) marks the heart of Leo the Lion and is the first-magnitude star closest to the ecliptic (thus a frequent target of Sun, Moon, and planets). It is about as far as Canopus and Aldebaran, and about as luminous as Aldebaran but far smaller than those other two stars. Its spectral class is close to that of Rigel.

Boötes
(Hu Gadarn)

The constellation of Boötes dominates the spring sky; it is a relatively large constellation to which the eye is immediately drawn due the presence of the bright star Arcturus. Boötes depicts the figure of The Herdsman or shepherd, guarding the flocks of his master and seeing off any danger that may threaten. To aid him in this activity, Boötes on many old star maps is seen to be holding a leash that is attached to the collars of the hunting dogs Canes Venatici. These constellations are supposedly chasing away the two celestial bears.

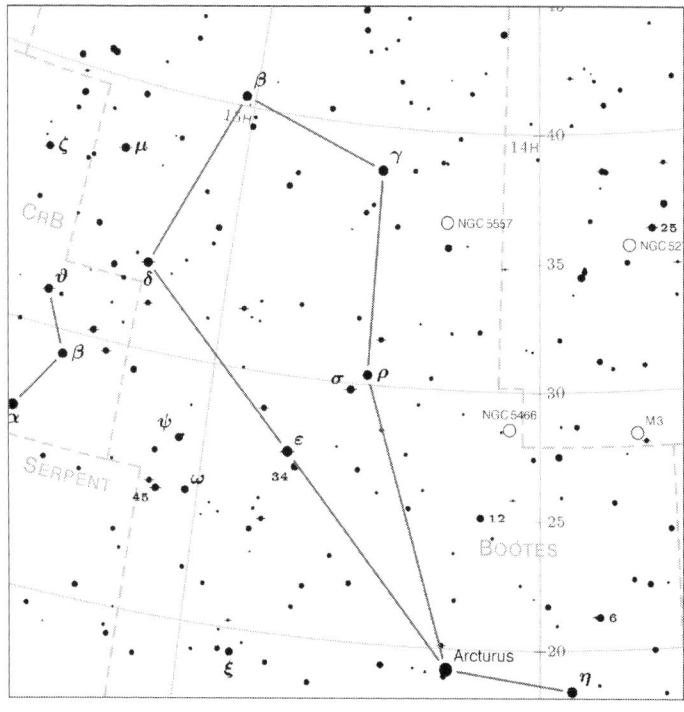

Boötes

In the tales of Wales, Boötes is the hero Hu Gadarn, the first person to link oxen to the plough. Hu is portrayed as holding the handle of the plough as the blade cleaves the sky, attached as it is through the yoke (Auriga) to Taurus the bull. This line-up of the constellations acted as a crude agricultural calendar, reminding the Celts to plough and sow the fields when seen in this configuration after sunset.

Hu Gadarn is also involved in another Welsh tale, that of the Addanc (associated with Scorpius). Hu is called upon to rescue the land from a fearful monster which is terrorizing villages and flooding the farmland. He uses his team of oxen to draw the Addanc out of Llyn Llion and drag it to Llyn y Ffynnon Las, where it is magically imprisoned.

* * *

Apart from the proximity of Arcturus, Boötes is a rather bland constellation for the deep sky observer, containing no star clusters or nebulae. Even the few galaxies that inhabit this region are too faint to be seen with modest equipment. Interestingly, cosmologists have discovered that the largest void, or space between the galaxy clusters, lies in the region of Boötes, so it may not be too surprising that it lacks a few redeeming features.

As Boötes is another constellation where one is looking out into the halo of our Milky Way, and indeed into deep space, it is no surprise to find that it contains just one globular cluster. This is NGC 5466, RA 14h 05m 30s Dec: 28°32m, a ninth magnitude object in the northwest of the constellation. It is a fairly distant globular, over 52,000 light years away, but is rather unremarkable in a small telescope, in which it is merely a fuzz of light, with the nebulous look of unresolved stars.

Arcturus is a truly amazing star. Its name means 'the guardian of the bear', which considering the herdsman's role, is strange, but its nearness to the circumpolar constellations of the bears must have led the ancient Arabian astronomers to associate it with these groups. It is the closest star of the group known as halo stars or Population II objects and is a K type giant about 25 times the diameter of the Sun. This is admittedly rather small when compared to stars such as Betelguese or Antares, but as Arcturus lies only 36 light years away so at this distance it is bound to outshine even these celestial searchlights. At present, Arcturus is at closest approach to our Sun, and will disappear from naked eye view in 500,000 years as its proper motion takes it at high speed towards the constellation of Virgo.

Another star of interest is the double star Mirak, or ε Boötes. It is one of the closest double stars that can be resolved with a small telescope, but good conditions and steady seeing are required to detect it. The primary star is a warm orange colour that can be seen even in binoculars, but the companion is an eighth magnitude speck of blue light, very close to the primary. It is a stiff test of the resolving power of your telescope, so if you can detect it, then you know that your particular telescope was indeed a wise selection. Boötes contains little else except for the star T.Boötis, a flare-like variable, which has only ever been observed once! It lies in the same low power field as Arcturus, but is usually below 17th magnitude, so it is far too faint for exploration with amateur equipment.

Canes Venatici
(The Cwn Wybir and the Cwn Annwn)

Dogs for hunting were a necessary accompaniment to any great estate and it is known from ancient records in Wales that freemen were obliged to maintain the king's hunting dogs and horses during his tours, a custom which obliged tenants to keep pack hounds for their landlords.

In the mythology of our dark land, the stars were originally part of Berecynthia, though some tales also have Hu Gadarn holding the reins of two hunting dogs as they chase a mythical bear around the pole. Canes Venatici could perhaps represent the Cwn Wybir or the Sky Dogs of which it was said that it was bad luck to hear them as they presaged the death of a family member.

The story entitled *An Account of Apparitions of Spirits in the County of Monmouth,* by the Rev E. Jones, states "The nearer these dogs are to a man, the less their voice is, and the farther the louder, and sometimes, like the voice of a great hound, or like that of a blood hound, a deep hollow voice." They are also known as the Cwn Annwn, the Dogs of Hell, in most tales and are described as being white with red ears and flashing eyes bright with fire.

* * *

Canes Venatici is a rather obscure constellation composed of faint stars, two of which are easily visible to the naked eye lying below the tail of Ursa Major. The group is meant to represent the two hunting dogs of Boötes, Asterion and Chara, names that were recognized by the Chinese where the dogs were assigned the role of guardians of the emperor's heir. This reveals that the constellation must have been a separate part of the circumpolar stars in the past, though the Greeks and Romans thought of the group as part of Ursa Major. Hevelius gave the group its modern interpretation. Canes Venatici is especially rich in all manner of deep sky objects, of which many can be seen with a pair of binoculars or a small telescope.

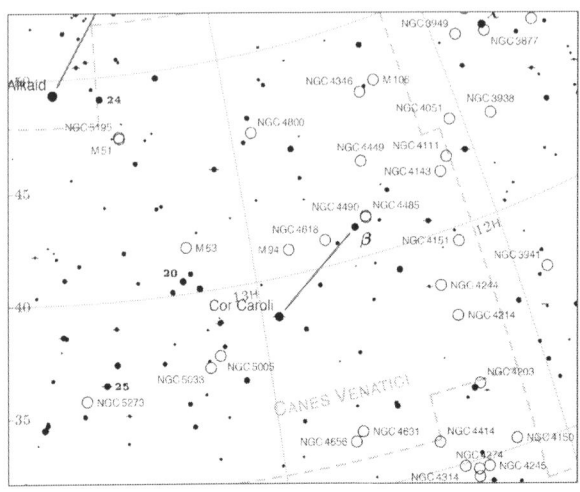

Canes Venatici

The most magnificent globular cluster after M13 can be found in the southern part of Canes Venatici, namely M3, a huge aggregation of stars 34,000 light years away. M3 is slightly easier to resolve in a small telescope than M13 and its starry points burst forth in profusion from this beautiful object, and some outliers may even be resolved with a good pair of binoculars. M3 contains upwards of 1 million stars, and the highest number of RR Lyra type variables yet found in a globular cluster. These variables are a fast type of Cepheid, going through their light amplitudes in a matter of hours rather than days.

Messier 3

Two galaxies of note in Canes Venatici are both Messier objects. The first, and easiest to spot is M94, an E1 type galaxy that lies slightly north of and in the centre of a line drawn from α to β Canum. It looks like a fuzzy star in binoculars, but a small telescope will only reveal a slightly larger bright white smudge of light with no detail detectable. The other galaxy to be seen is M63, which lies north of α Canum, and is a ninth magnitude smudge of elongated light that reveals itself to large telescopes as an Sb type spiral. Both galaxies are reputed to be outlying members of the Virgo group, approximately 55 million light years away.

One of the most beautiful galaxies in the heavens is found in the north of the constellation. This is the incomparable M51, the Whirlpool Nebulae, as Lord Rosse called it. It is a face-on Sc type, which is interacting with an irregular type galaxy. This interaction has given rise to many knots of bright blue stars, and has distorted the shape of this exquisite object. M51 is a ninth magnitude galaxy possibly the only object of this type where the spiral arms can actually be seen, winding around the nucleus in a glorious array of light.

Messier 51

Close by is the enigmatic object known as M106, a strange galaxy lying in the northwestern portion of the constellation. In a small telescope, a very bright nucleus can be seen, plus a condensation that reveals itself as a spiral arm. This galaxy is undergoing some form of explosive activity, similar to the Seyfert galaxies and has been classified as a Sa pec type. Another galaxy of note is the wonderful NGC 4362 in the south of Canes Venatici. In a small telescope this galaxy can be seen as a sliver of electric blue light, this colouration arising from the fact that this is another active galaxy, possibly arising from a recent collision with a near neighbour. Canes Venatici is a fruitful hunting ground for galaxy observers.

Coma Berenices
(Nad the Maiden)

Long hair was a feature of the ancient Celtic world and neither male or female would cut their hair unless something drastic had happened. An example of this occurs in a tale from North Wales, that of Llyn Nad y Forwyn or the Lake of Nad the Maiden. The legend recounts how a young man and a girl were walking by a lake on the evening before their wedding. Unknown to the maiden the man loved another woman and was planning to murder her. As they strolled along he suddenly pushed her into the lake where she drowned. However, her spirit troubled the lake and the surrounding area, and her ghost would occasionally appear as a ball of fire or a lady dressed in silken clothes walking the same sad pathway that brought her to her death. At times, the maiden would emerge from the waters, half naked, with dishevelled hair that covered her shoulders which she tore off in clumps and threw to the winds. The frequently appearing ghost with her hair blowing in the wind or running with water was taken by the Tylwyth Teg and placed in the sky as a permanent reminder of the young man's misdeeds.

This group is home to some of the most fascinating objects that can be seen with modest instruments. Coma Berenices was named by the Greeks in a charming legend regarding the loyal wife of Ptolemy, governor of Egypt. Berenice promised the gods that she would offer her shining tresses if her husband returned home unscathed after battle. Ptolemy duly returned in good health, and Berenice kept her promise. The glittering tresses disappeared however, and Solon, the court astrologer, was called upon to explain who had stolen them. In a laudable bout of quick thinking, he pointed out the constellation to the King and Queen and told them the gods were so pleased with such a sacrifice that they had placed the offering in the heavens. Berenice was placated, and the fearful astrologer, no doubt breathing a sigh of relief, lived to a ripe old age.

* * *

Coma Berenices, which is also known as Melotte 111, is actually one of the nearest star clusters to Earth, lying about 280 light years away. It is an outstanding sight in binoculars, the field being filled with shining points of light. One of the most interesting things about the cluster is its deficiency of low mass stars and dwarfs. The reason for this scarcity of low mass stars is not known, but the cluster may have lost many of its stars through gravitational effects, in essence, throwing them out of the cluster if they came too close to their more massive neighbours. Comae contains 37 stars that actually belong to the cluster, others are merely background objects. The brightest stars within the cluster have a luminosity of about 50 times that of the Sun, but there are no giant stars within this sparse cluster. Spectroscopic measurements show that the brightest members are only now beginning to evolve toward the red giant stage, so the cluster may be upwards of 70 million years old.

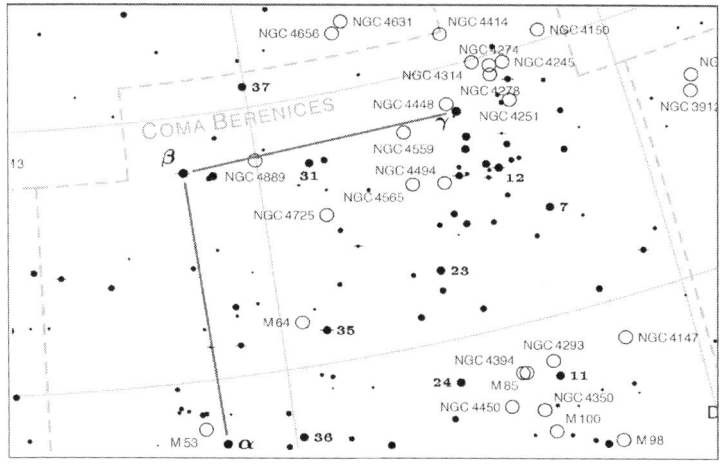

Coma Berenices

The brightest deep sky object in Coma the globular cluster M53, which is a magnitude eight object to be found only a degree or so away from α Comae. It is visible as a hazy patch in binoculars, but a small telescope will begin to resolve the outer stars into tiny needle-like points of light. M53 is 60,000 light years away, and is close to the faint and depleted, NGC 5053, for which you need a fairly large aperture to appreciate this dull citizen of the sky.

The constellation is within Hubble's wonderful Realm of the Nebulae, and as such many galaxies proliferate. It is impossible to grasp the nature of this region of the sky with the intellect alone; the realm of the nebulae a place where the sheer overwhelming depth and size of space begin to take a hold and the imagination is freed.

One of the brightest objects in the Realm of the Nebulae is the galaxy M64, commonly known as the Black Eye Galaxy for reasons which becomes clear as soon as you observe it through a telescope. The disc is partially covered by a huge cloud of dust and gas blocking out the light from countless stars, giving the appearance of a bruise similar to the arc of blue under a blackened eye. M64 is an Sb type galaxy lying 25 million light years away, and is thus one of the nearer members of the Virgo cluster. It is also the only Messier object discovered from Wales! First seen by Edward Pigott in 1781 from Frampton House near Llantwit Major, this little object made it into the Messier catalogue after being reported by Johann Bode, but the rights of discovery go to Piggott as he saw the object three weeks before Bode's report.

Messier 64

Other relatives of this galaxy teem in the spaces between the stars of Comae, and also abound outside of the star cluster itself, in fact the outsiders are easier to find! The biggest spiral galaxy of the whole Virgo group can be found in Comae, namely M100. This appears as a tenth magnitude smudge of white light through a telescope, but is difficult to see in binoculars other than a giant pair. M100 is 40 million light years distant and is an Sc type spiral that has a very high proportion of blue giant stars than is typical for such an object.

Another showpiece galaxy is the face-on Sc type galaxy M99, which can be found fairly close to M100. Although these galaxies bear Messier numbers, Pierre Mechain was the actual discoverer, Messier adding them to his catalogue after Mechain had brought them to his attention. M99 shows itself as an almost circular blob of tenth magnitude light, this shape apparently due to its face-on orientation. One fascinating aspect of M99 is that astronomers have discovered it has the largest Red Shift of any of the Virgo galaxies, and thus may be an interloper. Its distance is approximately 50 million light years, but may be a little higher.

Without doubt, the most astounding deep sky object within Comae is the edge on galaxy NGC 4565 at RA 12h 36m 18s Dec 25°59m. This sliver of tenth magnitude light lies close to the star 17 Comae, and reveals an exquisite lane of dust dividing the galaxy into two halves, a small telescope will uncover this remarkable feature, and the galaxy will withstand high powers that will show a little more detail, especially the tiny star like nucleus and the spindle of light that form the arms of this incredible object. NGC 4565 lies 20 million light years away, rather closer to us than the rest of the Virgo cluster, but it could be an outlying member, just as our local group is considered to be a part of this large aggregation of star cities.

NGC 4565

Within the bounds of Coma Berenices is the central area of the Virgo cluster, a fact only discovered earlier this century. The centre is marked by the proximity of a bright galaxy called M88, an elongated smudge of blue white light shining at tenth magnitude. The similarity in the magnitudes of all these objects is positive proof that they lie at approximately similar distances, and that they are comparable in their light output and therefore the numbers of stars inhabiting each one can be roughly gauged. M88 lies at a distance of 41 million light years and is one of the few Virgo cluster galaxies to show a hint of detail, even in a small scope. Additionally, although its quoted magnitude in most catalogues is magnitude 10.5, it appears to shine rather brighter than this and will take a higher power well.

There are countless other galaxies to look out for in this exceedingly rich area of sky. In some instances, a wide field ocular will show up to half a dozen galaxies in the field of view making identification very difficult. As the area is not well endowed with bright stars it can also be difficult for amateurs to find what they are looking for by 'star hopping'. Many of these galaxies have undergone supernova activity in recent years, so it can be profitable to know this area well, consequently identifying any newcomers at once.

Corona Borealis
(Caer Arianrhod)

In Welsh mythology this constellation, the northern crown, represents the goddess Arianrhod, the daughter of Don and the sister of Gwydion and Caswallawn. She joined Caswallawn in some latter dated tales to chase Julius Caesar out of Wales.

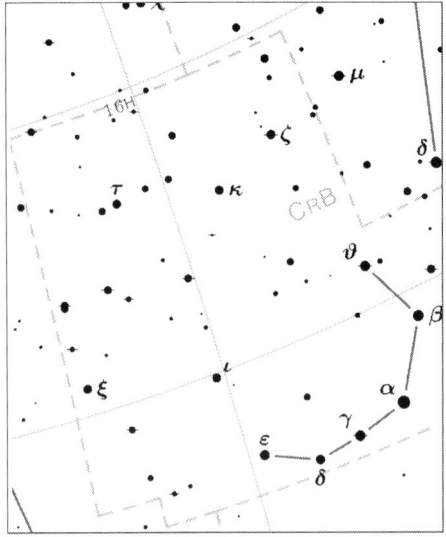

Corona Borrealis

The circlet of stars was known as Caer Arianrhod and is connected to a coastal rock formation in North Wales known as Arianrhod's Palace, which is only visible at low tide off the coast near Conwy. Arianrhod could take many forms and in Welsh tradition she also becomes the enchantress Cerridwen in the *Mabinogion*. She was the mother of the poet Taliesin in early Welsh tales and was apparently one manifestation of the inventor of the wheel – a useful thing for agricultural carts.

* * *

The main attraction of Corona Borealis is without question the variable star R. Corona Borealis, but this is a variable with a difference as R Coronae usually stays at maximum light, around 6th magnitude, before fading away, sometimes down to 13th magnitude. R Coronae is a late type M or R type object that has intense lines of carbon in its spectrum. As such giant stars give out large stellar winds, what probably happens is that R Coronae is hiding behind vast clouds of 'soot' that condense as the material of the stellar wind cools and becomes grainy.

The light fluctuations are irregular in period; it is not possible to predict when R will disappear, so it is worth keeping an eye on it whenever it is visible. R Coronae is the prototype of this kind of variable, and is a halo object, a population 2 star lying between 2500 and 4000 light years away, the discrepancy in the computed distance lying in the fact that different observations of the star have revealed different spectral peculiarities that give rise to opposing computed distances. Whatever the case, R Coronae is the most attractive and wondrous star of its type and can be easily seen with binoculars or a small telescope.

α Coronae or 'Gemma' is a double star that unfortunately can only be split with a spectroscope, but close by is a field of faint stars that contains the only well observed recurrent novae. The star is T. Coronae, which appears to go through novae-like eruptions every 80 to 100 years, the last time in 1946. This period is by no means fixed however, so scan the field and identify this 10th magnitude star so that you can observe it regularly.

The star ϱ Coronae has an attendant planetary system, which was discovered in 1997. The planet is another massive Jupiter and is extremely close to the parent star, only 0.23 AU from the star and having a mass of 1.5 times that of Jupiter, orbiting the star in 40 days. The orbit is so close that the atmosphere of this planet is estimated to be in excess of 300 degrees and may be blowing off into space with the impact of the stellar wind from the star.

Corvus
(Bran the Giant)

Historical associations with this constellation are rife across the Celtic world. None can be as powerfully epitomized as the legend of Bendigeidfran (the Blessed Crow King) or Bran the Giant from Welsh folk tales. Bran is the brother of Branwen and together they are the son and daughter of the sea god Llyr. In an attempt to unite the Celtic kingdoms of Wales and Ireland, Branwen was married to Maddolwch, the King of Ireland. This liaison should have brought peace but Bran's half brother Efnysien destroyed the trust of the Irish when he hamstrung the horses Bran provided as the dowry of Branwen.

The Irish sailed away, the king with his new bride, but despite the promises of Bran to send more horses, and in immediate recompense giving Maddolwch a magical cauldron that can bring the dead back to life, the Irish king was insulted. He took the counsel of his wise men who recommended that he could not break the peace, but could mistreat his bride. Despite the mistreatment Branwen gave birth to a son, Gwern, though this did not mollify the king. Branwen suffered all sorts of misfortunes until she tamed a starling that enabled her to get word to Bran that all was not well. The tales carried to Bran angered him and he raised an army and a fleet of ships to carry them to kill the Irish king and take back his sister.

The armies clashed outside Dublin. The fighting was hard and bloody and Branwen's son Gwern was killed by his half uncle Efnysien, before Efnysien himself died destroying the magic cauldron that was bringing the Irish dead back to life. Only seven Welshmen survived the battle but Maddolwch was defeated, though Bran was mortally wounded by a poisoned arrow. As his body was too large to be brought back by the remaining ships of the fleet, he asked his followers to cut off his head and then bury it in London, with his face to the south to stop the land being invaded. In Celtic afterlife beliefs, souls become birds; Bran's soul became a raven and his burial place, Bran's tumulus, is now the site of the Tower of London. Tradition states if the ravens desert the Tower, the land will fall to foreign invaders, giving us an entertaining story and direct link back to Bran's raven as the ancient Britons knew it.

* * *

Corvus is a rather dim constellation comprised of four main stars that make up a trapezoid found above the back of Hydra. Corvi is the brightest

The Spring Constellations

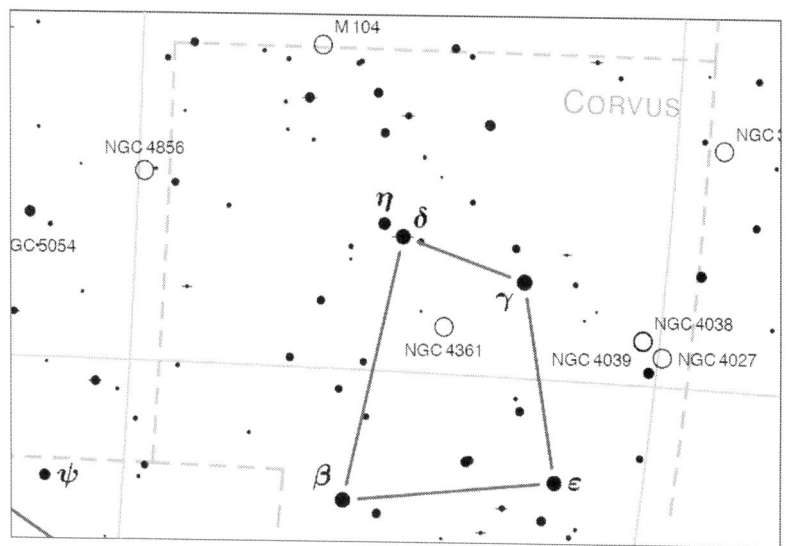

Corvus

recognizable star in this constellation, and marks the upper right hand corner of the trapezium. δ Corvi is a wonderful coloured double, conspicuous in a small telescope. The primary is a brilliant white in colour whilst the 8th magnitude companion is orange.

One of the most famous deep sky objects in the heavens lies right on the northern border of Corvus. This is NGC 4594, the Sombrero Galaxy, as it looks like the renowned Mexican hat in a small telescope. The object is otherwise known as M104, although Messier himself did not record it, but it is something of a surprise that he did not see it, as it is a fairly conspicuous galaxy shining at ninth magnitude.

A good pair of binoculars should pick up this fuzz of light against the background stars, but the view through a modest telescope is amazing, the whole galaxy is edge-on to us and is bisected by a dark dust lane that extends into the darkness of the night on its southern side. NGC 4594 is part of the Virgo cluster but is removed from the central area by many degrees. Its distance is estimated to be in the region of 29 million light years away.

Within the trapezoid of Corvus there lies a large, faint planetary nebula which, although it is of 10th magnitude, the surface brightness is rather low, so it does not present a dazzling target for modest equipment. The nebula is known as NGC 4361, and may be captured photographically. The only other

NGC 4594

object worth hunting for is NGC 4038, a pair of colliding galaxies lying about 90 million light years away. The fascinating aspect of these galaxies is that the collision has led to the creation of long streamers of stars radiating out into space, which has given rise to the name the Ringtail Galaxy for this object. Unfortunately, this kind of detail is beyond the range of amateur equipment, although a small telescope on a good night may reveal the 11th magnitude blob of grey light that constitutes this fascinating object.

Corvus contains many faint galaxies, but these are beyond the range of average instruments. However, the magnificence of the Sombrero galaxy more than makes up for the inadequacy of this charming constellation.

Crater
(The Cauldron)

The constellation of Crater – The Cup – was introduced by the Greeks, and is meant to represent the cup that contains nectar, the drink of the gods, which Hebe the cupbearer brought in daily to Zeus. Hebe eventually fell out of favour with the king of the gods, but why the cup was placed in the sky is a mystery in Welsh legend however because cups or cauldrons feature in so many tales. In some it is the Holy Grail, the cup used by Joseph of Arimathea to catch the blood of Christ.

Welsh tales from medieval times claim Joseph took the cup to Britain and founded a line of secret guardians to keep it safe. In modern times it came to

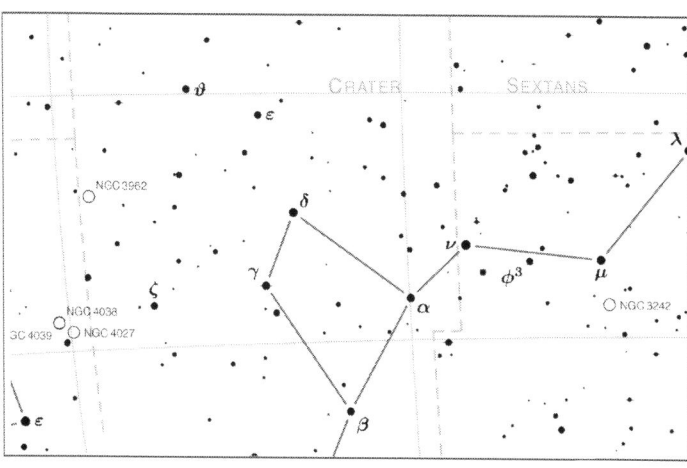

Crater

rest at Plas Nanteos a mansion near Aberystwyth, and over the years has attracted many visitors who drank from it or touched it, believing that it had the power to heal all ills. Today, the cup is a shell of its former self, with parts bitten off by the sick in the hope of a miracle cure. Nonetheless, belief in the cup's powers have persisted, despite a TV documentary in which experts found it dated some 1,400 years after the death of Christ. In 2014 the cup was stolen and there is still an ongoing police investigation but most have us have to look no further than the heavens where the constellation of Crater reminds us of the Holy Grail.

There is also an old Welsh story called *The Spoils of Annwn*, which tells of a war-like invasion of Annwn or the Welsh Otherworld by Arthur, whose target was a magical cauldron, described as made of shining bronze and studded with precious stones. This cauldron is probably the same as the one known as Pair Drynog and is one of the thirteen treasures of ancient Britain collected by Merlin and hidden away. This cauldron needed the breath of nine virgins to heat the broth within it and it would never feed a coward. Arthur's expedition ended in a less than glorious victory for him and his men: he gained the vessel but lost most of his men to the forces of darkness in doing so.

* * *

Crater contains not a single deep sky object that is within reach of average amateur equipment, although about a dozen galaxies are within the range of a 250mm reflector. It does have a number of double stars that are fairly unremarkable objects with little separation between the components. The brightest of these objects is the star γ Crateris, a 4th magnitude star which is a white primary accompanied by a ninth magnitude star which is also white, thus making a rather unspectacular pair.

Even for the variable star observer, the pickings are rather sparse. The brightest variable is R. Crateris, an M type red giant which varies irregularly between magnitude 8 and magnitude 9.5 in a rough period of 160+ days. Additionally, the star U Crateris can be followed through its cycle by telescopes of 150mm or more. The star fades from 9th magnitude to magnitude 13.5 in a period of 305 days.

The brightest galaxy in Crater that can be glimpsed in good seeing conditions is the Sc type spiral NGC 3887 which has a visual magnitude of 11.6, so it is not an impressive object. In all, Crater may well be a constellation that you would pass through without giving it a thought. It is to be found west of Corvus, lying above the back of Hydra. To observers with binoculars, this is a constellation of despair, although scanning the field stars is always worthwhile, as one never knows in advance if a nova or supernova could occur here. The constellation lies just outside the Milky Way, and is therefore viewed through the halo of our galaxy.

Hercules
(The origin of the Celts)

Hercules is of course the hero of legend, the strongman who carried out the twelve labours. Hercules was the son of Zeus, but was hated by Hera, the wife of Zeus. As a result, Hercules suffered unjustly for most of his life, but was rewarded with a place in the heavens and immortal heavenly life. The origins of Hercules as a Celtic deity is interesting. According to the Greek poet, Parthenius, the Celts themselves were descendants of Hercules (or Heracles to Greeks). When Hercules travelled back to Greece with the cattle of Geryon after the tenth of his twelve labours, Celtine, the daughter of Bretannus, fell in love with the famous strong man. One day, she hid the cattle, and would not tell Heracles where they were hidden until he slept with her. Hercules did so and Celtine became the mother of Celtus, the ancestor of the Celts.

In Celtic myths Hercules is renowned for other things, chiefly being a giant, and in Welsh mythology there are traditionally four: Idris, Ysgydion, Offrwm, and Ysbryn. Hercules is also associated with fertility gods as represented in

The Spring Constellations

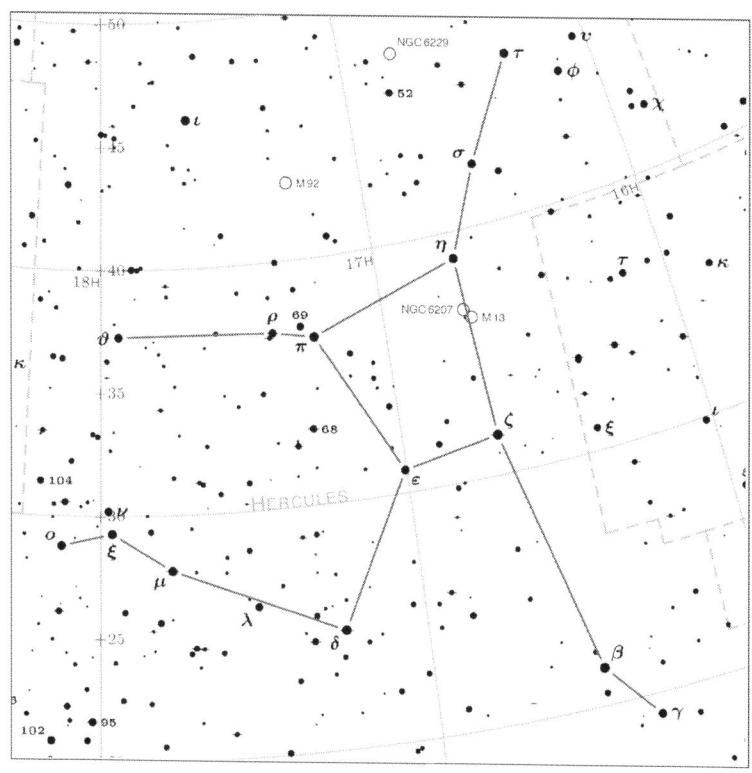

Hercules

chalk figures such as the long man of Willmington and the giant at Cerne Abbas where the giant is known as Helith. Anthropologists claim a connection between Helith and Hercules, the figure being an earthly depiction of this god with his club. The Cerne giant is dominated by its phallus; a symbol alluding to fertility rites going back to ancient times. Indeed, this phallic tradition has a modern counterpart. Atop the hill above the figure is an area where maypoles were erected to celebrate the coming of spring, the productivity and fertility of the land and the awakening of nature.

Hercules has been associated with various Welsh giants, the most prominent being Idris Gawr, who was a king of mid Wales in the early medieval period. Idris would sit upon a mountain top and survey his whole kingdom. Naturally, this mountain is Cadair Idris, (Cader Idris in some translations) the Chair of Idris', near Dolgellau in southern Snowdonia. The other three giants of tradition, Ysgydion, Offrwm and Ysbryn are also said to have mountain seats in the vicinity of Cadair Idris, though they may be part of the same massif. Looking at the steep northern side of Cadair Idris, there are three prominent peaks before one comes to the highest peak of all which is known as Penygader or the 'top of the chair'.

Cadair Idris is also the hunting ground of the Cwn Annwn controlled by Gwyn ap Nudd who is part of the legend of Gemini – the fighters over the lady Creiddylad, so it is easy to see how intertwined some legends are in their association between land and sky.

* * *

Hercules is one of the faintest of the major constellations in the spring sky, most of its stars being around 3rd to 4th magnitude. Nevertheless it presents an easily recognizable figure that dominates a large area of sky between Coronae Borealis and Lyra, and is a pleasing counterpoint to the straggling group of Ophiuchus to the south.

The figure of this constellation is outlined by six stars of which the upper four form a trapezoid known as the Keystone. Within its bounds can be found one of the most beautiful objects in the northern sky; the great globular cluster M13. This object was originally discovered by Edmund Halley, but was noted by Messier later in the century. His comment describing M13 as a "round nebula containing no stars" must be the greatest mistake of his observing life, and points to the quality of his telescope, as even a 60mm refractor will resolve some stars in this lustrous object.

Messier 13

Through a telescope, M13 looks like a woolly ball radiating tiny pinpricks of light in all directions, with high powers on a large telescope the cluster takes on a brilliant three dimensional effect, with the centre beginning to show resolution into numerous stars, and chains of starry light spreading out from the edges of this incredible object. A count of the stellar images on photographic plates show that the cluster contains at least one million stars, mostly of old red population 2 types, many of which have masses equal to or less than our Sun. M13 lies over 23,000 light years away from us in the halo of our galaxy.

In the early years of the twentieth century the American astronomer Harlow Shapley made an extensive study of globular clusters and found that most of them lay in a spherical 'shell' around the galactic nucleus, and were constantly orbiting our Milky Way, sometimes even passing through the galactic disc. It is currently thought that globular clusters were formed at the same time as our galaxy condensed from the gas clouds left over after the 'big bang'. Indeed, theories have been put forward that galaxies firstly condensed from gas clouds of globular cluster size, and that the remnants of this creative act are the globular clusters, the leftovers of galaxy formation.

Another showpiece object in Hercules is the globular cluster M92, which can be found easily in a pair of binoculars as it lies between the 'arms' of the figure (or even its legs as many think the constellation is portrayed upside down!). M92 lies at a considerable distance, over 26,000 light years away, but a small telescope should show some hint of resolution into stars.

On the opposite side of the constellation can be found another globular cluster NGC 6229, which is a ninth magnitude blob of light and does not reveal any spectacular form in a small telescope. Hercules also contains one planetary nebula that can be seen in binoculars or a small telescope, but it is a rather difficult object to distinguish. This is the planetary NGC 6210, RA16h 44m 30s Dec: 23°49m, a 9.5 magnitude object in the southern part of the constellation, which is very star-like, its disc being almost imperceptible in modest instruments. It shines as a bluish, out of focus star in an area of sky devoid of bright stars, and consequently is not difficult to find.

The major star of Hercules is α Herculis, otherwise known by the name of Rasalgethi, which means the 'head of the kneeler' (which alludes to the constellation being upside down). This is one of the most beautiful stars in the sky as Rasalgethi has a green companion of 5th magnitude, which is a little difficult to spot in a small telescope, the separation being only 4.6", but is worth the effort to find.

Rasalgethi lies over 360 light years away by current estimations, but may actually be more distant than that. It is a halo star, but is not a population 2 object, rather it is a red giant of spectral type M5II, and is one of the largest known stars. It is at least 400 times the diameter of the Sun, and its surface temperature is a cool 2500 degrees Kelvin. So cool is α Herculis that the bulk of its radiative output is in the infrared part of the spectrum. If our eyes were sensitive to this wavelength of light, then Rasalgethi would be the brightest star in the heavens, with Mu Cephei and Betelgeuse coming a close second.

Hercules contains a very distant cluster of galaxies, but these are not the types of objects that will interest observers with modest equipment, as the brightest of them does not reach 14th magnitude. However, there is one galaxy visible to observers with a six-inch telescope and above. This is the lovely NGC 6207, only half a degree away from M13, and can be seen in the same low power field. It is an Sb type galaxy, possibly an outlying member of the Virgo group as its distance roughly corresponds to the known distance of this cluster of galaxies at 45 million light years.

A lovely planetary nebula for a small telescope is NGC 6210, a green disclike object forming a triangle with two stars in the southern part of the constellation near the western shoulder. A small telescope will reveal the green disc and its fuzzy, unfocused outline quite clearly. NGC 6210 is 6500 light years away and at that distance is over 1.5 light years in diameter.

Hydra
(The Gwiber of Penmachno)

In mythology, the Hydra is a snake associated in Babylonian records with waters and fountains, and is the sign of the goddess Tiamat. In Greek mythology, Hercules killed this legendary monster as one of his twelve labours. In some old Welsh tales, the Hydra is associated with a mythical dragon as its sinuous body curves across the sky. However, we know that Draco is also seen as a dragon and there may be some confusion between this constellation and that of Hydra. It probably represents the tale of the Gwiber, a long snakelike beast with small wings that terrified the area around Penmachno.

A young man called Owain ap Gruffydd came to the area to kill the Gwiber but upon calling on a local man with the gift of foresight into the future, he was told that he would die in the attempt. Although he disguised himself three times on different days to consult the man, a different death was foretold for each disguise: Owain would die by the Gwiber's bite, by having his neck broken and by drowning. He laughed at all these pronouncements and revealed himself to the prophet, claiming that he would survive as the old man knew nothing of the future.

Owain sought out the Gwiber but its speed took him by surprise. He was wrapped in its coils before he could even draw his sword. The Gwiber then bit him, the venom flowing into his veins; shocked by the attack he fell over a tree stump and as the beast uncoiled itself he struck his head, breaking his neck. Finally, his body rolled into a nearby stream and powerless to do anything to save himself, he drowned.

The people of Penmachno found the body and, enraged that their champion was dead, set off after the Gwiber and filled its body with arrows and spear points, destroying its little wings in the process so it could not escape. In its death throes, it disappeared into the river whence it came and was never seen again, though it is said that the fairy folk were displeased at the Gwiber's death and placed it in the sky as a warning to mortals.

* * *

Hydra is the longest and most sinuous constellation in the whole sky. So long is it that its head is visible in winter, and its tail disappears mid-way through summer. Despite its length, Hydra is a constellation of faint stars, none of which rise to more than second magnitude. However, the head of Hydra is very distinctive, and once found, becomes the gateway to navigating the rest

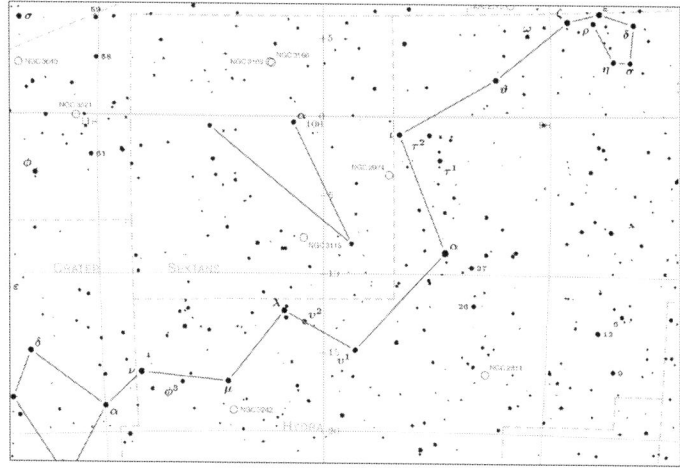

Hydra

of the constellation. Hydra lies under the constellations of Leo, Corvus, Crater and Virgo, but despite its association with such rich constellations, it does not have many deep sky objects of interest to the casual observer.

One of the finest deep sky wonders in Hydra is the star cluster M48 which is a large irregular cluster of 50 or so stars in a fairly rich field close to the Milky Way, and which lies adjacent to the constellation's border with Monoceros. M48 shines at magnitude 5.5; its stars are mostly A type giants which have a visual magnitude of 9th to 13th mag. The cluster lies 1500 light years away, and is famous in the Messier catalogue for its spurious position: it lies over 5 degrees away from the position Messier indicates. A beautiful sight in a small telescope or binoculars, and may even be visible to the naked eye on a night of exceptional 'seeing'.

Messier 48

Another deep sky object deserving of attention is also one of the most beautiful planetary nebulae in the night sky. This is NGC 3242 at RA 10h 24m 48s Dec -18°38m, an eighth magnitude disc of light that resembles the disc of a planet more than any other planetary in the sky. This object has on occasion been called the Ghost of Jupiter due to its uncanny resemblance. It lies over 2000 light years away, and is a marvellous sight in a small telescope, on high powers its lovely blue disc fills the field, and subtle shading indicates the presence of gaseous shells. This planetary can even be seen in binoculars on a good night, as a tiny star-like disc.

The real showpiece of Hydra is, sadly, not seen very well from Britain. This is the barred spiral galaxy M83, which shines at eighth magnitude, but due to its southerly declination it is extremely faint from these latitudes. It can be seen as an elongated blur of bluish light in a telescope, but binoculars may well provide a better view. M83 is relatively close by, only 10 million light years away, and may be an outlying member of our local group.

Just to the southeast of α Hydrae at coordinates RA 09h 34m 50s DEC -12 07m 46s is the 6.5 magnitude star HD 82943, a G0 type lying 70 light years away which contains two planets. The primary one has a mass of 1.6 times that of Jupiter and orbits at 1.16 AU from the star in 444 days whilst the other planet lies within this orbit at 0.7 AU at a mass of 80% that of Jupiter and orbits in 221 days. Such systems hold out the hope of finding ones similar to that of our own.

There are two globular clusters visible in Hydra, both rather unremarkable objects due to their southern aspect. One is M68, a bright globular which shows little detail in a small telescope, but can be discerned as a 7th magnitude smudge of white light above the tail of Hydra. The other object is NGC 5694, which at magnitude 11 is beyond the capacity of most amateur instruments. It is another globular of distinction, being very distant, and practically outside the confines of our galactic halo, on a par with NGC 2419 in Lynx. Hydra contains many double stars of note, but recourse to an album such as the Webb Deep Sky Society's Binary Star handbook is recommended.

Leo
(Twrch Trwyth and King Arthur)

In the Brecon Beacons there lies a valley known as Cwm Twrch. The name translates as 'Valley of the Boar' and perhaps derives from the Twrch Trwyth, a mythical wild boar of Arthurian legend of which is found in the ancient tales of the *Mabinogion*. There are several variations this tale. Culhwch wishes the hand of Olwen in marriage but her father Ysbyddaden the giant is reluctant to permit his daughter to wed for he knows that when she marries, he will die. Ysbyddaden sets Culhwch several near-impossible tasks which, with the enlisted help of Arthur and his knights, are accomplished one by one, including the hunting down of the boar by Mabon and the hunting dog Drudwyn.

In the *Mabinogion,* one of King Arthur's tasks was to rid the western Brecon Beacons of the pack of wild boars that were terrorizing the people. He chased the boars from Dyfed eastward towards Powys. It was on the Black Mountain that he picked up a large stone and hurled it at the pack, killing its leader on the edge of the valley near Craig-y-Fran Gorge. The big boar's body rolled down the valley and into the river, now known as the Afon Twrch. The big stone, known as Carreg Fryn Fras is still on the mountain. Geologists have a different explanation for its origins, as it is probably an erratic. The river forms the boundary between Carmarthenshire and Powys, anciently the kingdoms of Deheubarth and Brycheiniog.

* * *

The constellation is a wonderful asterism of stars that actually does resemble the animal it is meant to feature. The bright star Regulus, or α Leonis, lies right on the path of the ecliptic, and can therefore be occulted by the Moon and even the planets during certain periods. This is one of the closest of the bright stars as it is positioned about 79 light years away and has a luminosity of 160 times that of our Sun. Regulus is an B spectral type star, and appears distinctly bluish to the naked eye. The star also has a companion visible in a small telescope, an 8th magnitude star that is a pale yellow in colour.

If the observer follows the 'sickle' up towards the top of the mane, then you will encounter one of the loveliest binary systems in the spring sky. The star is γ Leonis, the brightest star of this part of the mane; it can be seen well in a small telescope, the companion star is of 3rd magnitude and is fairly close to the primary. The stars are both yellow in colour, but various observers have described the stars as orange or even greenish!

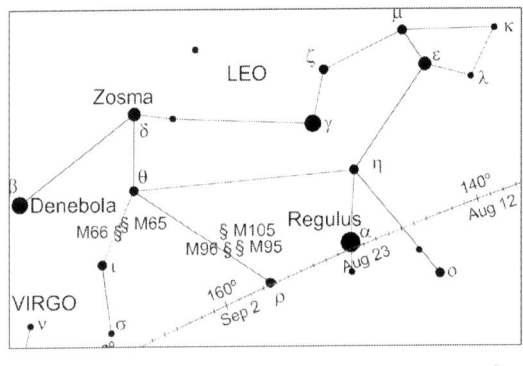

Leo

One of the most extraordinary stars in the sky can be found in this constellation. R Leonis is a long period variable of the Mira type that is one of the easiest stars of this class to find, as it makes a little triangle with 18 and 19 Leonis. The star fluctuates between magnitude 5 and 10 in a period of 312 days. The whole period can thus be easily followed with a pair of binoculars and is a worthy long term project, as accurate records for long period variables are few. R. Leonis is approximately 600 light years away, and is one of the most beautiful stars coloured stars in the sky. At minima, it has been described as a 'fiery-red' with a hint of purple.

Leo is found in that part of the sky where we are looking out into deep space rather than at our own galaxy. Therefore it is only to be expected that we will find some bright galaxies to examine with binoculars or a small telescope. The first such galaxies that must be visited are the bright pair known as M65 and M66, located just below the body of the lion. Messier's friend Pierre Mechain first noticed these two spectacular objects in 1773. Observers with a pair of binoculars can see the pair as a faint smear of light lying close together in a field of sparse stars, whereas a telescope will show a little more detail. These galaxies are spirals of type Sb, and if you look closely at M65, you may be able to discern a thick lane of dust bisecting the arms of this object.

Leo triplet

Just to the north of these galaxies can be found the fainter NGC 3628, which is an 11th magnitude sliver of light. Unfortunately, it cannot be seen in binoculars, but a small telescope may well reveal it. These three galaxies appear to be outlying members of the Virgo cluster, the closest of the great galaxy clusters to our Milky Way. Their distance is therefore in the region of 35 million light years, making M65 and M66 quite luminous objects.

Leo contains another twin set of galaxies to explore, although both may be difficult objects to resolve. M95 and M96 can be seen together in a low power eyepiece, but they are unlike the bright M65 and M66. Again, Mechain discovered both objects. M95 is one of the closest examples of a barred spiral galaxy, lying 38 million light years away, whilst its companion M96 is a Sa type galaxy at a similar distance.

A bright galaxy that merits the observer's attention is located just in front of the mane of Leo, below the star λ Leonis. This is the bright galaxy NGC 2903, an Sc type spiral galaxy at a distance of 31 million light years. It shines as a 9th magnitude object that should be faintly visible in a good pair of binoculars, and shows up well as a blue smudge of light in a small telescope.

Leo contains very little else of interest to amateurs with minimal equipment, but the constellation has an additional grace in that it is the radiant for a meteor shower, which on one occasion, held the population of the western world spellbound as over 100,000 shooting stars rained down into the Earth's atmosphere. This happened in 1833, and again in 1966. Although the shower occurs every year in November, the cycles of brilliant showers appear to come every 33 years, so 2032 will be a year to look forward to. This 33-year interval is due to the orbit of the parent comet responsible for the Leonid stream. Comet Tempel-Tuttle is a faint object with a period of 33.17 years, the orbit of which sharply intersects the Earth. Several other meteor showers are visible during the year, the most famous and brightest being that of the Perseids, which occur during August. Reference to a star atlas such as Nortons' will give the times of these events and a few others of note.

Observers with larger telescopes will find Leo a good hunting ground for galaxies. What could be more relaxing than finding these faint objects and contemplating the gulfs that lies between us, yet allowing the imagination to soar to these majestic islands of stars that populate the fantastic universe we inhabit?

Virgo
(The Lady of Llyn y Fan Fach)

This constellation is meant to represent the eternal virgin, the goddess Ceres, the patron of agriculture. However, in Celtic mythlogy she is associated with healing and is the goddess Bridgid or Briget. This name was taken by an early Christian nun who looked after the sick and injured and has been canonized as St Bridget or as we know her in Wales, St Bride. There are many churches dedicated to her and old holy enclosures with the epithet Llansantffraed.

Due to the association with healing, the constellation can be linked with one of the most enduring local legends, the Lady of Llyn y Fan Fach. This famous lake in the eastern part of the Brecon Beacons National Park is a favourite destination for walkers, the cool waters of the lake surrounded by the crags of Bannau Brycheiniog and the Carmarthen fans.

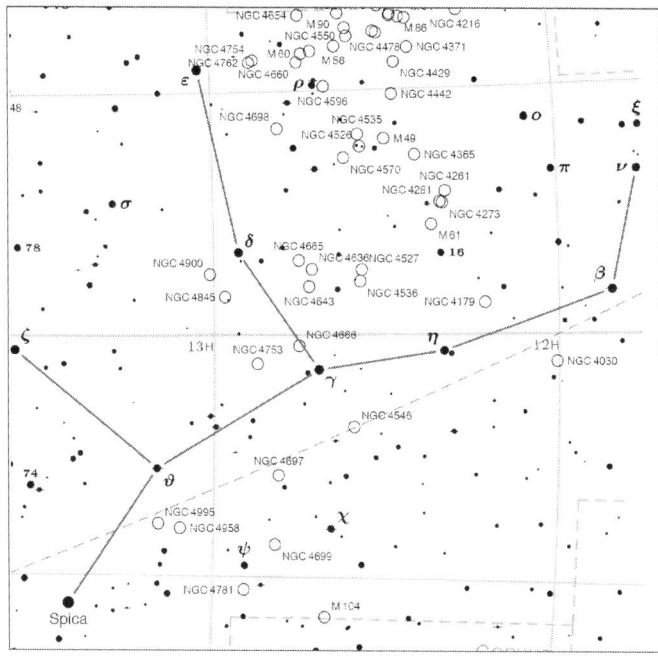

Virgo

Legend has it that a local widow's son by the name of Gwyn saw a beautiful woman bathing in the lake, or rather standing on the water arranging her hair and using the smooth surface as a mirror. He fell in love with her, and her father assented to the marriage as long as the young man did not break a particular vow — the lady would return to the lake if her struck her three times. The father also gave a dowry of beautiful white cattle and Gwyn and the lady settled down to married life and had many sons. During the marriage, the white cattle became numerous and much sought after, spreading eventually across mid Wales. The descendants of these famous cows can be seen at Dinefwr Park outside Llandeilo.

However, as in many Welsh tales, Gwyn lost his wife after he playfully slapped her with some gloves, then again after finding her crying at a christening and for the third and final time when he found her laughing at a funeral. The lady took her dowry and returned to the lake and Gwyn died jumping into the lake in an attempt to recall his wife. Nevertheless, the lady did not forget her sons and they were taught at the lakeside by her in the arts of healing and making potions. She showed them a sacred place, Pant y Meddygon, (physician's dingle) where grew healing herbs and special plants. The boys became the famed Physicians of Myddfai and the Lord Rhys Crug gave them oversight of the lands around Myddfai. One of the famous Myddfai physicians even became a doctor to Elizabeth I. The lady of the lake and her healing powers are commemorated in the starry outline of the heavenly lady.

* * *

Virgo describes a broad Y shape with the first magnitude star Spica at the base of the Y. The bowl of the Y then faces the constellation of Leo, or Twrch Trwyth. Within the boundaries of the bowl can be found one of the greatest aggregations of galaxies visible from Earth. The Virgo galaxy cluster is the closest such association to our solar system hence it is bursting with detail. In modest amateur telescopes, upward of 50 Galaxies can be identified in the group; the only problem in seeing them arises from the confusing profusion of these objects. The Virgo group lies at an approximate distance of 53 million light years, and contains all known classes of galaxy, including some very active ones.

Markarian's chain

On photographs taken by the world's largest telescopes, over 3000 such objects have been counted in an area of sky measuring 10 x 12 degrees, corresponding to a linear diameter of 5 million light years for the group. The cluster congregates around a collection of galaxies commonly known as Markarian's Chain. The galaxy cluster itself may well be an outlying member of the Virgo-Centaurus galaxy cluster, an aggregation of star cities that number into the tens of thousands.

One of the most studied galaxies of the Virgo group is M87, an E0 type galaxy of 8th magnitude discovered by Messier in 1779 whilst observing a comet, which passed through this intriguing region. M87 can be seen as a small fuzz of light, almost like a faint globular cluster in either binoculars or a small telescope. It is the most massive galaxy known to man, containing at least 3,000 billion stars and over 1000 globular clusters. It is also an active galaxy, being the radio source Virgo A; additionally, it is a bright source of x-rays and possibly cosmic rays. It can be located a few degrees away from the star ρ Virginis.

Messier 87

Two excellent galaxies for small telescopes lie close by M87. These are M84 and M86 respectively. They can be seen together in a low power eyepiece and are both elliptical galaxies of ninth magnitude. In the same low power field can be seen two other galaxies under good seeing conditions, these are NGC 4438 and NGC 4435, both are around magnitude 11.5. M84 and M86 are an ideal starting point to discover the rest of the galaxy cluster as they mark the brightest part of the centre of the cluster.

The brightest galaxy of the whole group for amateurs is the lovely M49, an E3 type elliptical galaxy of eighth magnitude that can be seen a round smudge of light close to two 6th magnitude stars. It is about 6 degrees removed from the centre of the cluster but is relatively easy to find. In binoculars it can be seen as a faint mist of light, and good eyesight is essential to pick out this object with such equipment. M49 is another massive system, apparently having a mass of five times that of the Milky Way. M49 is 56 million light years away.

A galaxy of exceptional interest is M59, a barred spiral type which shines at tenth magnitude. In a moderate-sized telescope the central bar can be resolved, with the arms making a misty blur of light surrounding the bar. If you have an eight inch telescope, then try spotting a pair of interacting galaxies in the same low power field about half a degree south of M59. These are NGC 4567 and NGC 4568, otherwise known as the Siamese Twins as they appear to be joined together at one end. They are both 12th magnitude objects and are Sb type spirals.

A rewarding object to discover is M61, a face-on Sc type spiral shining at tenth magnitude. It is visible in a small telescope as a ghostly pool of grey light, which in a larger telescope resolves itself into beautiful arms with a star-like nucleus. It lies 52 million light years away. Interestingly, several supernovae have occurred in recent years within this galaxy, so it is worth following every night it is visible, just in case another one goes up. A further bright spiral worth noting is M90, an Sb type lying 60 million light years away. It can be seen in binoculars as a faint bluish patch of luminosity, but a telescope reveals it to be a large almost face-on spiral very similar in size and composition to our Milky Way. It too is characterised by an intense star like nucleus.

Virgo also contains one of the most interesting binary systems that can be followed with a small telescope or even binoculars if they are properly mounted. This is the star Gamma Virginis; otherwise know as Porrima or Vindemiatrix. The star is easily located at the bottom of the Bowl of Virgo, and both components are yellow stars of the same apparent luminosity, shining at magnitude 3.5. The system has a period of about 180 years and is closing to periastron at present. This closest approach happened in 2007, and the components have been drawing out ever since so are a little easier to separate.

A good star atlas, planetarium programme or app is a vital piece of equipment enabling one to explore this fascinating constellation further. The proliferation of bright galaxies is a sight not to be missed, and if you are interested in supernova patrol work, then this is the place to look if you own a reasonably sized telescope. However, all observers can revel in this beautiful part of the heavens and contemplate the wonders found therein.

CHAPTER FIVE
The Summer Constellations

'Twas noontide of summer
And mid-time of night;
And stars, in their orbits,
Shone pale, thro' the light
Of the brighter, cold moon,
'Mid planets her slaves,
Herself in the Heavens,
Her beam on the waves.
I gazed awhile
On her cold smile;
Too cold - too cold for me -
There pass'd, as a shroud,
A fleecy cloud,
And I turned away to thee,
Proud Evening Star,
In thy glory afar,
And dearer thy beam shall be;
For joy to my heart
Is the proud part
Thou bearest in Heaven at night,
And more I admire
Thy distant fire,
Than that colder, lowly lght.

—Edgar Allen Poe

Chapter Five –
The Summer Constellations

The lighter nights of spring slowly give way to the beautiful midnight blue of the summer night sky, bringing with it the exquisite constellations of Cygnus, Aquila, Lyra, Sagittarius, Scorpius and many more that delight and stun the observer.

A famous asterism of the summer sky and a brilliant guide to the other constellations of summer is the so-called 'Summer Triangle' composed of the three principal stars of Cygnus, Aquila and Lyra, namely, Deneb, Altair and Vega: three very bright stars that are an obvious indicator of the presence of sensational sights to come. The summer skies are so rich that even the smaller constellations hold many deep sky jewels; groups such as Scutum, which contains the brightest star clouds in the northern sky, Delphinus, holding one of the most distant globular clusters and Sagitta, which embraces two beautiful Messier objects, make the summer period wonderful.

The chief constellations follow the path of the Milky Way, the highest in the northern sky being the constellation of Cygnus with its bright star Deneb. Across the Milky Way to the west is Vega and the constellation of Lyra.

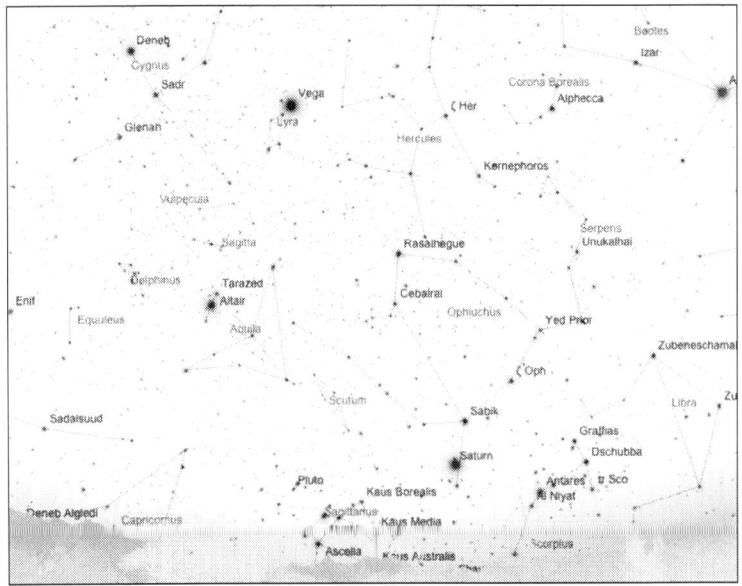

Summer constellations

Returning to the Milky Way, one descends from Cygnus past the indistinct stars of Vulpecula just off the beak of the swan past the little constellation of Sagitta and into Aquila. Just below Aquila on the Milky Way is the small group of Scutum Sobieski (Sobieski's Shield) and then one moves south into the magnificent constellation of Sagittarius, home of the galactic centre. Looking west from Sagittarius one comes to the top of Scorpius with its bright red supergiant star Antares and then returning northward up the Milky Way from here one encounters the bell-shaped Ophiuchus and the snake Serpens (Caput and Cauda) completing the round of the summer groups.

We are examining a large portion of the Milky Way galaxy in which we live, but in contrast to the winter sky, the stars of summer are relatively distant and are therefore slightly weaker in apparent brightness than the winter stars. The vast expanse of the summer Milky Way more than makes up for this deficit, however.

Sometimes however, the summer nights can turn cold, so it is best not to observe in tee-shirt and jeans, always make sure that one is properly dressed and then you can savour the summer sky all night long if you wish to. The grandeur of the sublime star fields of the Milky Way never fail to impress anyone who views them from the advantage of a dark sky site, trying to convey the wonder of this vision to others is summed up in beautiful poem by the Chinese sage Tao Yuan Ming:

> Somewhere, there lies a deeper meaning,
> I would like to say it,
> But have forgotten the words.

Distinctive Stars

The bright star Vega (Spectral type A0V) is directly overhead on a late July evening and is the most obvious star of northern summer. Vega's true brightness (as measured in absolute magnitude or in luminosity) and its true size are not much greater than those of Sirius. Vega appears less luminous than Sirius only because it is about three times more distant. The basic similarity between the two stars is their composition, as revealed to us by their spectral class – to which the colour and surface temperature correspond closely. Many naked-eye observers detect a tinge of blue in the otherwise whitish light of both Sirius and Vega, but you should record what you think of their colour on various nights

Vega differs significantly from Sirius in having no known true companion star, but it does have a physically unrelated star that happens to lie along almost the same line of sight and make a pretty view for amateur observers. Both of these false companions are about magnitude 9.5, one of them about 1 arc-minute south of Vega, the other about 2 arc-minutes northeast.

Altair, (spectral type A7V) located in Aquila the Eagle, is connected with Vega in the giant asterism known as the Summer Triangle. This star is 'only' 16.5 light years away, the fourth closest of the first-magnitude stars. It happens to be approaching us slightly more quickly than any of those other stars. Furthermore, a feature of its spectrum indicates that it must be the fastest spinning of the brightest stars. Whereas our Sun's mean rotation period is about 25 days, the equatorial rotation period of the larger Altair is estimated to be just 6.5 hours! This fast rotation is thought to deform the shape of Altair into a flattened ellipsoid, with the equatorial diameter being nearly twice the polar diameter. Whenever observing Altair, consider this amazing fact. Also, note the star's colour on various nights in various optical instruments. Finally, you may wish to note that Altair has a tenth-magnitude star, physically unrelated to it, glinting a little less than 3 arc-minutes to the northwest of it.

Antares (spectral type M5Ib) is one of the two first-magnitude stars in the sky that is a red giant. Although inferior to winter's Betelgeuse in brightness, size, and variability, this star marking the heart of Scorpius the Scorpion is prodigious nonetheless. If Antares is 550 light-years away, its true brightness would be about 8,000 times as great as our Sun's and its diameter 880 times as great. Another fascinating facet of the Antares system is its companion star, which looks vividly green in contrast to Antares itself. This companion can be seen in 150mm aperture and occasionally in smaller telescopes, but only when 'seeing' is excellent.

Deneb (spectral type A2Ia) the final member of the Summer Triangle, may have the greatest true brightness of any first-magnitude star, but is also almost certainly the most distant. If Deneb is 2600 light years away, then its light output is 80,000 times the Sun's. Although not one of the 0 or B class stars of very high temperature and typically great luminosity, Deneb is a prodigious object, containing perhaps 25 times the mass of our Sun. Deneb and Altair form a striking example of the extremes of star distance and star brightness among the first-magnitude class: Deneb appears fainter, but only because it is about a thousand times farther away. Indeed, an intelligent being on a planet near Deneb would need at least a 150mm telescope just to glimpse our Sun as a 13th magnitude star.

The Summer Milky Way

From the UK the Milky Way in summer presents a glorious arc of silver light that stretches from the northern to southern horizon and going right through the zenith. This presents an ideal opportunity to engage with our galactic home and with nothing more than a pair of binoculars, drift along noting all manner of objects.

The summer Milky Way is very distinctive and bright as we are looking into the whole of our galaxy the Milky Way from our solar vantage point. It is easy to see why the ancients thought of it as a river of milk as the white shimmering stream really does seem to pour down from the sky. In mythology the Milky Way was created when Hera, the queen of the gods and wife of Zeus found an abandoned child (Hercules) near the temple of Zeus in Olympia. Knowing of his fondness for human women and the possibility that this was one of his progeny, she picked up the child and at her daughter Aphrodite's insistence, fed the child by holding it to her breast. Hercules, the strongest man on Earth was also quite muscular as a child and as he fed, the pressure became too much for Hera who screamed and tore the child away from her, causing her milk to spray across the sky to form the Milky Way.

In the ancient tales of Wales, the Milky Way is Sarn Gwydion, or Gwydion's Way after the uncle of Llew Llaw Gyffes who, after the death of Llew took the embers of a fire and ascending into the sky to seek out the soul of his nephew, he scattered the embers behind him so that he could find his way back to earth; the embers, still brightly burning, became the stars of the Milky Way. To most modern eyes the shining bright road looks like a trail of ashes, a milky deposit in the sky in which it is impossible to make out single stars, the whole amalgamating into a river of light from countless starry points at great distance from us. In modern Welsh parlance the Milky Way is known as Llwybr Llaethog, an almost literal translation.

Considering how light polluted many locations have become, the number of people who have never seen the Milky Way is increasing generation after generation. As the director of Brecon Beacons Observatory, I am constantly delighted by the wonder and awe of persons introduced to the glorious sight of the Milky Way for the first time, especially as the summer Milky Way is a glorious sight from the Brecon Beacons International Dark Sky Reserve. Many times I have heard the disappointment of someone prepared to pack away as the "clouds are moving in" only to have them watch in amazement (and part foolishness) when told "they are not clouds – that's the Milky Way"! Urban dwellers are often astonished at the brightness and richness of our galactic home.

* * *

Starting in the north with the constellation of Cassiopeia, the observer notices that the star clouds of the galaxy begin to become defined around the area where the knotty star cluster M52 is located. Here one is looking at the Perseus spiral arm of our galaxy. Moving southward, one can drift through Cepheus, noting the increasing number and density of stars in the region between δ Cephei and Deneb in Cygnus, with the milky glow of both IC 1396 and its central star cluster, through the small but bright star cluster M39, to the great diffuse cloud of NGC 7000, the North American nebulae. The observer has now moved on to the Cygnus spiral arm. Note that the stars decrease in number on the western border of this object, an effect that is easy to spot in the wide field of binoculars.

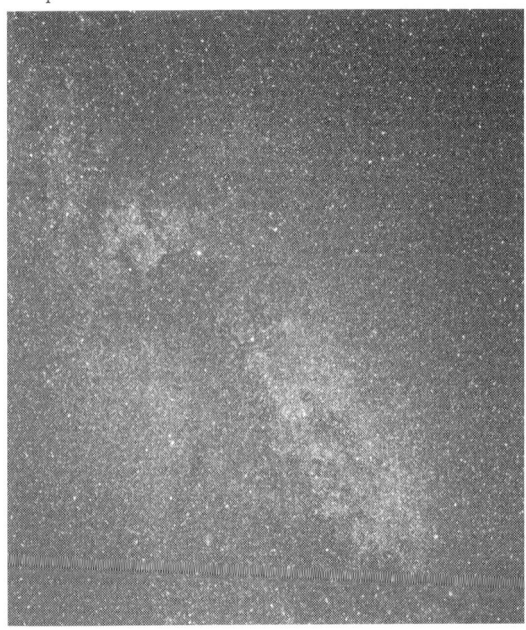

Milky Way

Just below Deneb, the Milky Way appears to split into two as the great Cygnus Rift, a large molecular cloud of interstellar gas and dust, begins to block the light from countless background stars. Scattered across this rift are the star clusters M29 and NGC 6871, neither of which are easily resolved, yet add to the journey with their sudden concentrations. Just below Albireo, the observer can look at Brochi's Cluster, otherwise known as the Coathanger. See how many of its stars you can count!

Here the Milky Way is noticeably divided with two main branches to east and west of the rift. As one continues into Aquila, the westernmost dark cloud spreads out and envelops the Milky Way, causing a complete break in the western section. At the northern end of this break, look for IC 4665, a little jewel of a cluster, located just above β Ophiuchi which shows to advantage in binoculars. Moving east from this across the stump of the western Milky Way, another lustrous cluster, IC 4756 can be observed. Going directly south, one encounters M11, the Wild Duck cluster and if one carefully examines the surrounding star fields, the dark clouds B111 and B119 can be seen flanking the cluster, whilst just further south and almost the size of the full moon, B112 is a lovely dark patch against the increasingly bright star clouds of Scutum, where the Crux-Scutum arm is seen at an oblique angle from our earthly perspective.

Scutum star cloud

Below the star cloud that marks the constellation of the Shield, one moves on to the Sagittarius spiral arm where the binocular observer can make out a plethora of star clusters and nebulae. Starting with the Eagle nebula M16, one hops through M17, the Omega or Swan nebula, through the star cloud M24 and a small cluster M18, southward to M21, and the nebulae M20 and M8 the Lagoon nebula. Tiny knotty-looking 'stars' in this area mark out the globular clusters NGC 6544 and NGC 6553 whilst to the east of them are the globular clusters M28 and the wonderful M22. Revisiting M8, the field is filled with star clouds differentiated by dark lanes, the largest of which is the Pipe Nebula or B59 to the west of M8 with clouds B65-67 ascending like smoke from the bowl of the pipe. Continuing westward into Scorpius, examine the area to the north and east of Antares, where the ϱ Ophiuchi complex lies as a darker mottling and decreasing number of Milky Way stars.

Milky Way centre

Returning to the centre of the Milky Way, which is noticeably bright here, the observer is now looking at the galactic centre, which is located four degrees off the western tip of the 'spout' of the teapot asterism making up the constellation of Sagittarius. Between the star γ Sagittarii and the galactic centre is a bright patch of stars about one square degree across known as Baade's Window. Named after the great twentieth century astronomer Walter Baade, this area gives the observer a direct view through the Sagittarius star clouds to the bulge of the galactic centre, a view that penetrates almost 25,000 light years into the heart of our galaxy. One should see a tiny knot of light – the globular cluster NGC 6522 that marks the centre of the window. The stars visible in this area are moving with a slightly different radial motion to the rest of the stars of the galaxy in this region, and have recently been found to be the stars of the central 'bar' of the Milky Way.

Look carefully around this wonderful area, which is crossed with dark clouds, bright star clusters and knotty condensations hinting at the arms and structure of our Milky Way. From the UK, the Milky Way here descends below our horizon, but why not reverse your course and go back up the river of light and discover how much more one can see.

Capturing the Milky Way is a magnificent way of starting out in astrophotography. If your telescope is equipped with an equatorial mount and is driven or is equipped with slow motion controls, you can make guided shots of the Milky Way with a little practise and patience, simply by piggybacking a SLR camera with a cable release atop the telescope. Remember to set the exposure timing to 'bulb' and exposures of tens of seconds to several minutes will bring out the best of the home galaxy. With a standard 50mm lens, you will not need to track too accurately, but keeping a bright guide star in the confines of a reticule within your eyepiece will enhance your precision, and make a pleasing photograph that can be enjoyed during rainy nights.

Aquila
(Gwalchmai)

In the *Mabinogion*, the constellation is known as Gwalchmai – the 'hawk of May' which was the afterlife form of the hero Llew Llaw Gyffes. Gwydion searched for the soul of Llew and found it as a hawklike bird. Retrieving the soul, he reanimated the body of Llew and they took vengeance on Goronwy and Blodeuwedd. The constellation is an alternate form of Llew.

The bird could also represent creatures known as the Adar Llwch Gwin, giant birds very similar to a Griffin that were given to the warrior Drudwas ap Tryffin by his wife, who was from the fairy world. These birds were said to understand human speech and to obey whatever command was given to them by their master, but Drudwas fell foul of them and they tore him to pieces.

Gwalchmai is also associated with Sir Gawain in Arthurian legends and is mentioned in many times in the *Mabinogion*.

* * *

α Aquilae, or Altair as it is also known, is the twelfth brightest star in the sky and is an A type star very similar to Sirius. It is also one of the nearer stellar bodies, relatively, lying as it does 16 light years away. Altair is of interest due to its very rapid rotation. It is a star with twice the diameter of the Sun, yet it completes one revolution in 6.5 hours, compared to the Sun's 25 days. As a result, Altair must be a very oblate spheroid, or even ellipsoid, although naturally such an effect cannot be seen with even the world's largest telescopes, let alone in an amateur instrument. Altair is easy to locate as it lies between two third magnitude stars in the centre of the 'bow'.

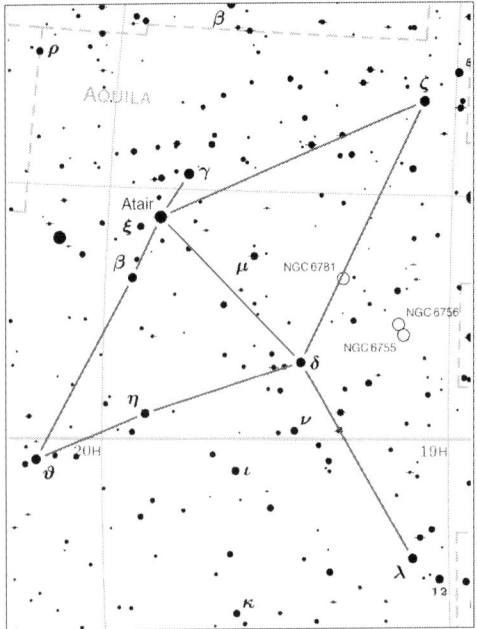

Aquila

Aquila contains a few star clusters of interest, one of the brightest of which is NGC 6709, RA 18h 51m 30s Dec: 10°21m, a group of forty or so bright stars in a lovely rich field. It is easy to locate in binoculars as its integrated magnitude is 8, but it is best seen with a telescope as it is a little too compressed for binoculars to separate the stars. An outstanding cluster worth hunting for along the star fields of the Milky Way is the wonderful NGC 6755, RA19h 07m 48s Dec 04°14m, a bright group of fifty stars in a large rich field shining at ninth magnitude, and easily visible in giant binoculars as a faint starry patch of the heavens. A small telescope will reveal some star chains radiating out of the rather loose centre.

The constellation also holds a small faint globular cluster that may just be seen on a good night with a small telescope. This is NGC 6760, RA 19h 11m 12s Dec 01°02m, an 11th magnitude blob of light that is difficult to discern against the Milky Way stars in an area of heavy obscuration. It is apparently a very rich, tight globular; the distance to the object is uncertain, but could be in the region of 20,000 light years, possibly more.

Aquila contains one of the highest numbers of planetary nebulae of any constellation. Unfortunately, many are fairly dim objects and are very difficult to distinguish against the background mottling of the stars. A few are outstanding however and are worth hunting down. One such is NGC 6781, a ghostly ring of light shining at 11th magnitude with a dimly lit interior. Look also for the green starry planetary nebula NGC 6790 which is also 11th magnitude but easier to spot due to its concentration.

A form of nebulae that can easily be distinguished from a dark sky site are the beautiful dark clouds that inhabit this area of our Milky Way in great profusion, due to the presence of the Cygnus Rift. These nebulae look almost like holes in the sky, and are best seen from a dark sky site with binoculars. Centre the field on γ Aquilae, and you should see two of the most obvious, B 143 and B 133. The first of these dark objects is a V-shaped patch obscuring the Milky Way about 1.5 degrees west of γ Aquilae. The other object, B 133, is shaped almost like a rugby ball and is a totally blank space with no stars marring its dark perfection. Both objects are best viewed when the 'seeing' is very steady, but ensure that your binoculars are mounted rigidly.

The closeness of the Milky Way to the bright stars of this constellation will enable the observer interested in the work of the nova patrol to be able to compare the magnitude of any newcomer with a fair degree of accuracy. Indeed, many novae in this constellation have been discovered with both binoculars and the naked eye. The last such naked eye occurrence was in 1918, when a formerly obscure star rose swiftly to magnitude -0.5, to become one of the brightest nova this century.

Novae occur where two stars orbit one another in a binary system. One of the stars is usually a main sequence object, whilst the companion is a white dwarf star. What astronomers have gleaned from observations of nova, and from computer simulations of these celestial fireworks, is that the dwarf star orbits its companion for many millions of years without anything happening. Eventually, the main sequence star shuts down its hydrogen burning, and swells to become a red giant. The former main sequence star becomes so large that it fills a domain within which the white dwarf now has some influence. The limit of this domain is known as the Roche lobe, the limit of gravitational control of the main sequence object. Once the Roche lobe has been filled by the expanding red giant, the star continues to grow in size and materials spill off the primary companion and through the inner Lagrangian point onto the white dwarf. Over time, the dwarf continues to strip off the enlarging envelope of its red giant companion and pull it onto its own surface, building up a concentration of gas over the surface of the dwarf. This gaseous matter does not burn immediately due to a phenomenon common to white dwarfs, called electron degeneracy. Although the dwarf's surface is hot, this degeneracy does not allow the white dwarf to burn off this layer of gas in thermonuclear reactions. Eventually enough material builds up on the surface of the dwarf until it is compacted by the gravitational pull of the dwarf star to such a state that it undergoes thermonuclear reactions on its own. In a brief instant, the whole surface layer converts into helium, giving out a tremendous amount of energy, brightening the star by up to 500,000 times the luminosity of the Sun, and enabling a formerly dim star to enjoy a moment of glory. This action does not destroy the white dwarf and material can build up again to explode another time. Astronomers are finding increasing evidence that the bright novae are not just one-outburst wonders. So, by keeping a watch on the heavens, you may be the next astronomer to spot a nova to equal the 1918 outburst. All you need is patience, dedication and luck.

Cygnus
(The Swan of St Bride)

This spectacular summer constellation is unmistakeable, lying along the Milky Way, seemingly diving down towards the rich star fields of the heart of our galaxy. Looking at Cygnus, it is easy to see why it has the alternative appellation the Northern Cross, as its six brightest stars do indeed form such a grouping. In classical Greek legend, the swan was placed in the sky by Zeus to honour the creature after he had transformed himself into one to enable the seduction of Leda, the queen of Sparta, thus fathering Castor and Pollux, the twins now called Gemini, and Helen of Troy and Clytemnestra, the wife of Agammenon.

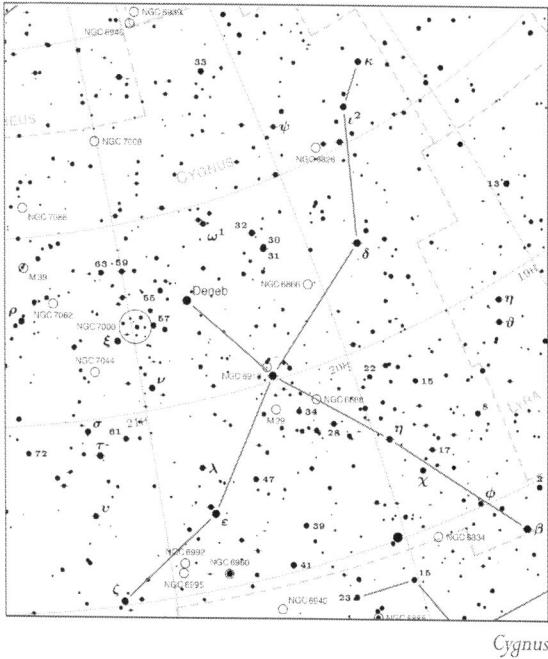

Cygnus

The constellation has some connection to Welsh literature and is often associated with the Irish saint Brigid or Bridget, who probably originates as a female deity within the iron age cultures of Britain. St Bride, or Bridget, could adopt a variety of animal forms, the most significant of which is a white swan. In her role as patron of childbirth and nursing, the goddess was also associated with the Milky Way, the river of light over which the celestial swan flies, and which looks back to the ancient Greek legends of the milk of Hera.

This same symbolism was associated by Welsh bards and druids with the goddess Minerva, a goddess of wisdom but here given the Gallic form of Brigantia, or Brigid, by the Romans. The druids were native to Anglesey and were eventually subjugated by the Romans. Druid means 'Knowledge of the Oak' and the potential between the wisdom of the Druids and the wisdom of

St Bride during the miracle of childbirth is sufficient to associate the swan of the sky with her.

* * *

Cygnus is crammed with deep sky objects for both the binocular observer and those with telescopes. It is without doubt one of the richest areas of the heavens for the British astronomer and more than one night is required if you wish to spot every object it has to offer. One that shouts the elegance of this constellation is the exquisite star β Cygni, or Albireo. This is the most incredible coloured binary system in the summer sky. The primary star is a golden yellow in colour whilst the companion star is a staggering electric blue, making a very pleasing contrast that no astronomer has yet tired of. Albireo is visible in good binoculars if they are mounted, and is definitely one of the showpieces of the summer night.

If you have access to a dark sky site, then scan to the east of the bright star Deneb, or α Cygni. There you may discern a faint lightness against the dark backdrop of night that is in reality the impressive 'North America' nebulae or NGC 7000, a beautiful object that can be captured photographically with a moderate exposure. This is one of the few objects that portray what their

NGC 7000

name suggests, although it is difficult to trace the full outline of this remarkable nebula with modest equipment. Its chief illumination is thought to come from the intense light of Deneb, one of nature's celestial searchlights, shining with a power of 80,000 times the light of our Sun and lying 1600 light years away. Interned within the nebulae is a faint but rich and crowded star cluster designated NGC 6997 that accommodates about fifty stars.

To the northeast of Deneb lies a beautiful open cluster that is easily seen in binoculars. This is M39, a group containing about thirty-five members shining with a combined luminosity of magnitude 6. Close by are two beautiful clusters for a small telescope, NGC 7082 and NGC 7062, lying among the stars of the Milky Way. NGC 7082 contains about forty stars in a compact, rich field and its stars are all magnitude 10 to 13 but blaze with a combined luminosity of magnitude 8. NGC 7062 is a smaller cluster in terms of content and size; it contains thirty stars of similar magnitudes to NGC 7082.

Messier 39

Travelling down the tail of the swan we arrive at γ Cygni, or Sadir as it is called. In the rich star fields to the southeast of this star can be found another cluster, a small compact group of twenty-five stars known as M29, which looks like a starry bundle through binoculars. Sweeping the area with such an instrument is very rewarding, as the Milky Way scintillates with the mottled luminescence of millions of unresolved stars, whilst here and there a bright knot betrays itself as a lovely cluster to entice the imagination. Around the area of γ Cygni are several red nebulous clouds of ionized hydrogen, which can only be seen to good effect by capturing them on film. They do however let themselves be captured relatively easily, and are a breath-taking sight.

Messier 29

Along the western wing of Cygnus are two incredible clusters that are not difficult to track down with a small telescope. They are NGC 6811 RA: 19h 38m Dec: 46°34m, and NGC 6866 at RA 20h 03m Dec 44° 0m. The first of these ravishing objects is a compact group of around seventy stars in a small area of space, very distant from us, yet glittering at 7th magnitude with the light of its white and blue giants.

Veil nebula

NGC 6866 is a little less interesting, but contains a nice crowded field of forty or so stars. Both clusters are 3900 light years away, and lie in a crowded field along the Milky Way. Continue along the western wing and you will soon find one of the most delightful planetary nebulae in the heavens. This is Herschel's Blinking Nebulae, or NGC 6826 at RA 20h 10m 24s Dec 46°28m as it is catalogued. It is an eleventh magnitude bubble of light that seems to appear, then disappear as soon as you look at it. Use averted vision to stabilise your view of this telescopic object and bring out its ethereal beauty. It lies in the same telescopic field of view as the bright double star 16 Cygni.

Underneath the western wing is a brilliant showpiece object, the star cluster NGC 6819, which borders on a seventh magnitude star. This incredible cluster is astounding in a small telescope, its 150 or so stars radiating at tenth magnitude in a rich but large cluster of stars that are at a distance of 7200 light years.

Passing over to the other side of the constellation, exploring underneath the eastern wing of Cygnus will uncover a fabulous network of gaseous filaments, a supernova remnant that must rank as the finest in the sky. This is the fabled Bridal Veil Nebula, the remains of a star that blew itself up over 50,000 years ago. It cannot be seen in binoculars, unless you have a giant pair, but in an 80mm telescope, the easterly portion known as NGC 6992 can be seen from a dark sky site as a ghostly grey-blue rainbow of light over 1 degree in length. Most people are drawn to the star 52 Cygni, simply because many observatory photographs show the nebulae, known as the 'Witches Broom' wrapped around this star. Through even a moderate telescope, the observer can discern the difference in the numbers of stars on each side of the nebulae, one side is packed with starlight and the other is simply empty space, or a huge cloud of obscuring dust. Apparently the Bridal Veil nebula lies about 1500 light years away and is 70 light years in diameter.

Above the eastern wing of the swan is one of the most famous stars in the sky. This is the rather dim 61 Cygni, a fifth magnitude star that lies fairly close to our solar system, only 11 light years away, and is in fact the fourth closest star to us. Its fame lies in the fact that it was the first star to have its distance accurately measured by F.W. Bessel in 1838 using the parallax method. 61 Cygni is one of the loveliest doubles, its companion is of 6th magnitude and is easily separated, although there is no colour contrast, both stars are K type orange dwarfs with a luminosity of about half of that of our Sun.

If you follow the Milky Way through this constellation with the naked eye, then you will discern that this stream of stars appears to divide into two parallel sections. This is the famous Cygnus Rift, an obscuring band of dust and gas that is characteristic of spiral galaxies like ours. NGC 4565 in Coma Berenices is an excellent example of what our galaxy would look like from the outside, whereas we are viewing this feature from the interior of our galactic home. You can follow the rift right through Aquila and Ophiuchus, right down to Sagittarius, where it becomes lost in the confusion of the galactic nucleus.

Several hundred variables are within the light grasp of the modestly equipped observer; the problem lies in identifying your target amongst the millions of background stars! One of the best variables in this constellation is χ Cygni, a red giant that varies from 4th or 5th magnitude to under 12th magnitude in 406 days. Another interesting variable is P. Cygni, although this is a variable at permanent maxima! P. Cygni flared up to third magnitude in the eighteenth century, but has settled down to 6th magnitude at present. It is considered to be a very young star that has shed its swaddling bands of dust and is now a main sequence member of the Milky Way. It must be a brilliant object as it is considered to lie over 7000 light years away.

A binary system of particular fascination is a pair of blue supergiant stars glowing at seventh magnitude near the star η Cygni. The stars are known by their catalogue designation HD 226868, and are visible in a small telescope as they are bright enough to stand out against the backdrop of the Milky Way. The curious thing about one of these stars is that it is orbited by a small, unseen companion that is the x-ray source Cygnus X1 and has been confirmed as a black hole, perhaps the only such system within the grasp of amateur equipment. Obviously, no telescope in the world will reveal the presence of the black hole, but it is interesting to consider this star and its life history, and ponder on the significance of this outstanding object.

If you have recourse to a good star atlas, there are countless more deep sky wonders that merit consideration, whilst the binocular observer may savour hours of delight by slowly scanning this fantastic, crowded region of our galactic abode. The presence of the Milky Way is an open invitation to nova and supernova hunters, and it is worthwhile getting to know the constellation well in order to identify any interlopers. On average, there is at least one nova a year in this rich region, but most stay below naked eye magnitude, although one never knows when a bright one like Nova Cygni 1975 may flare up.

Delphinus
(The Dolphin)

In Celtic times, the dolphin in the sky was an important constellation as the creature was recognized as a good luck charm. Perhaps its habit of swimming alongside small boats as the ancient Welsh fished the rich waters around the coast made the mammal good company for the fishermen, and their playful antics possibly reminded them of their intelligence and friendliness. Seeing a dolphin was meant to bring good luck to the watcher and also signified that good weather was guaranteed – perhaps this is why the dolphin is a summer constellation.

This lovely little constellation is easily envisaged as the dolphin it is meant to represent. Apparently, it is the dolphin that came to the rescue of the famous Greek poet Arion, who was returning by ship from a competition in which he had won a large sum of money. The avaricious sailors robbed him of the prize and cast him overboard, hoping to hide their crime. But Zeus sent a dolphin to rescue him, and then destroyed the ship with a thunderbolt. Arion was saved and the dolphin thereafter was placed in the sky.

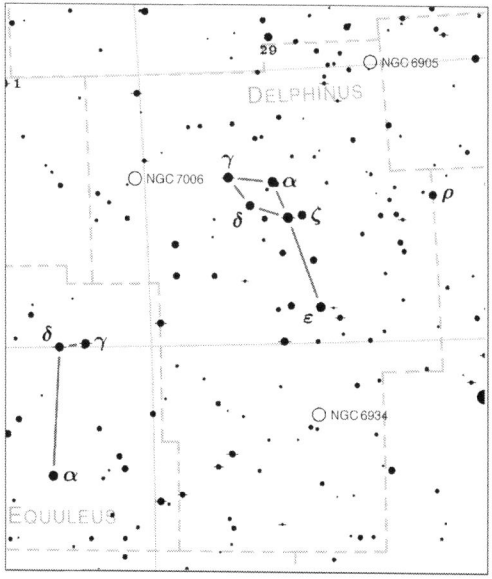

Delphinus

A tale informs us that the constellation may be associated with the mermaid of Cemaes Head. A fisherman from St Dogmaels hauled in an extraordinary catch one day and as he cleared his nets he could see that he had trapped a beautiful young woman with the lower body of a fish and a large tail. He thought that there would be a good price for the creature so he began to row to shore as quickly as he could to show off his prize.

The mermaid begged him in a low and sorrowful voice to let her go – she could not live out of water for long and she was failing, with her scales cracking and their beautiful colours fading. She promised that if she was released she would save his life one day. Sad that the creature was dying and remorseful that he would be the cause, he let the young mermaid go and watched her disappear into the waves.

Many years later he was about to set out fishing in his boat when he saw a head crowned with golden hair emerge from the waters in front of him. He recognized the mermaid straight away and she told him to turn back to shore this instant. He warned the other fishermen but they ignored him and went out to sea. The minute they were out of sight of Cemaes Head, a storm suddenly swept upon them and many lost their lives. The fisherman was saved by the warning of the Mermaid of Cemaes Head but many of his companions that would not heed his words were gone forever.

* * *

The trapezoid shape of the body of the dolphin contains the lovely coloured double star γ Delphini, a superb sight through a telescope. The primary is blue-white in colour whilst the companion is a glowing yellow, which has sometimes been described as pale green, due to the contrast with the primary. It lies relatively close to our solar system at 101 light years.

Delphinus contains two globular clusters that are worthy of amateur attention, although it must be stated that neither object is particularly bright or detailed. The first of these objects is the ninth magnitude NGC 6934, RA: 20h 34m 12s Dec 07°24m; a small round bundle of white light lying directly below the tail of the dolphin in a rich field of the Milky Way. It should be well seen in good binoculars, whilst a small telescope will show it to be slightly mottled with many field stars seemingly making a ragged 'resolved' edge of pinpoints to the cluster. NGC 6934 is approximately 52,000 light years away.

The real celebrity of Delphinus is the very distant cluster NGC 7006, an object on a par with its distant cousin NGC 2419 in Lynx. The globular NGC 7006 shines feebly at 11th magnitude and is a rather small, difficult object to capture against the Milky Way star fields, but from a dark sky site, with a small telescope, you should be able to pick up this elusive ball of light. NGC 7006 is very distant, over 185,000 light years from our solar system, making the separation between it and NGC 2419 to be over 300,000 light years. It is not known if NGC 7006 is actually a member of the Milky Way cloud of globular clusters, or if it is intergalactic tramp. It has a luminosity of 130,000 times that of the Sun and possibly contains more than 80,000 stars.

Lying amongst the stars in the northern part of the constellation is the planetary nebula NGC 6905 a beautiful vivid blue little nebula with an elongated axis. It is known as the Blue Flash Nebula due to its wonderful colour and is located 3000 light years away, and yet is quite large at that distance, subtending 1.2 arc-minutes. It can be found at RA 200h 22m 23s Dec 20o 06m 18s.

Delphinus accommodates a few variable stars of interest to the observer the best of which is R. Delphini, a Mira type long period variable, which alternates between mag 7.7 and 13.7 in 264 days, it can therefore be followed part way in binoculars. Delphinus is also a good hunting ground for nova observers, the bright nova HR. Delphini becoming a remarkable naked eye object in 1967, after its discovery (with binoculars) by the brilliant British amateur observer G.E.D. Alcock. This discovery was a boon to professional observers, as HR. Delphini remained bright for over one year after discovery, the slowest fading nova on record. So never give up: increase your knowledge of star patterns and fields to increase your chances of unearthing this beautiful type of object.

In the northern part of this constellation, amongst the star fields of the Milky Way is an interesting extrasolar planetary system orbiting the star HD 195019, a GIV type sub giant at a distance of 75 light years and shining at magnitude 6.9. The coordinates for the system are RA 20h 28m 17s Dec +18 46m 12.s. The planet is at least three times the mass of Jupiter and orbits at a distance of 0.14 AU in just over 18 days. Although the star is bright enough to be seen in binoculars, its position should make it a rather challenging object to find.

Libra
(The Flooded Field)

The constellation of Libra is one of the less populated regions of the summer sky, and if it were not for its position on the ecliptic, enabling some planets to be seen in the group from time to time, it would be written off as far as modestly equipped amateurs are concerned. Libra was once a part of the large constellation of Scorpius to its east and is associated with the Welsh monster known as the Addanc. On old star maps it originally depicted the outstretched claws of the creature, struggling to come to grips with its enemy, Orion, far to the west. As this would make the astrological zodiac into eleven signs, the Romans added Libra to the zodiac in 46 BC, an act later eulogised by Milton, who in *Paradise Lost* attributed this 'balancing' of the zodiac and the calendar to God:

> The eternal, to prevent such horrid fray
> Hung forth in heavens his golden scales, yet seen
> Betwixt Astrea and the scorpions sign

In Welsh mythology the constellation is meant to represent a wide flood of water and is known as Lli Bras or Llyn Bras, a reference to the damming power of the constellation of Scorpius to which it is closely related in Welsh tales as that constellation is known as the Addanc or beaver. Indeed, the constellation lies in front of the monstrous creature as if it is going to disappear from human sight at any moment into the waters of its own making. The name Llyn Bras refers to the lake that such a creature could almost conjure up overnight and the frustration of the local farmers can be imagined to find a perfect field with ripening crops vanish overnight in a flood.

Flooded land is a feature of ancient Welsh mythology, and appears in the story of Cantre'r Gwaelod, a land that stretched across Cardigan Bay from Ramsey Island in the south to Bardsey Island in the north. The area was flooded when one of two princes, sons of Gwyddno Garanhir, left open the sluice gates that kept out the sea after drunken carousing. The sea cascaded in and the land has been flooded ever since. It has been discovered that around the tenth Millennium BC, Britain was joined to Europe by an area now called Doggerland located under the North Sea and that the coasts of Wales were further west than they are today, so perhaps this tale is a cultural memory of the land now lost to the waves.

* * *

The most interesting star within the constellation is β Librae, or Zuben El Genubi as the Arab astrologers named it. This title means the Northern Claw, another reference to the fact this group was once part of Scorpius. The interest lies in the colour of this star, as several observers have noted it as the only obviously green coloured star to the naked eye in the whole sky. β Librae is a B type star lying at a distance of 140 light years. Usually such stars are a blue-white in colour, but the low elevation of the constellation as seen from Britain may account for the reports of its greenish hue, as observers can easily note how the star scintillates in the rising atmospheric currents of summer.

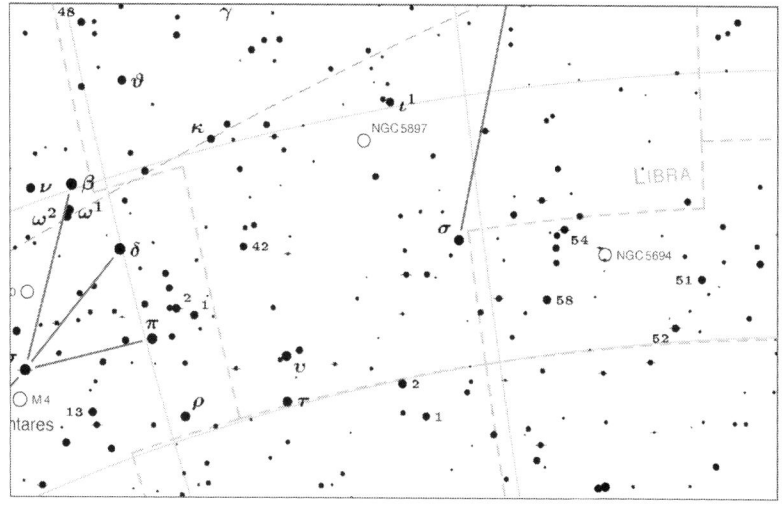

Libra

Libra contains one globular cluster that is available for scrutiny by the observers with small telescopes. This is NGC 5897, RA 15h 17m 24s Dec -21°01m, a tenth magnitude smudge of light that is irresolvable in such instruments. The globular itself is not a spectacular object compared to some of its brothers and sisters, being rather loose and depleted in stars. It is an outlying member of the group of clusters currently closing on the nucleus of our galaxy in their orbits, and is approximately 35,000 light years away. It may just be seen as a round blob of light in binoculars under exceptional seeing conditions.

δ Librae is an eclipsing variable of the Algol type, but the magnitude range is not great, only just over 1 magnitude. It is, though, a useful object to observe which will train you to gauge stellar magnitudes accurately. The period is 2.34 days, and the star is 225 light years distant.

Libra is a good hunting ground for observers interested in galaxies or equipped with instruments of large aperture. Most of the galaxies are below

12th magnitude; about a dozen are visible within the constellation. It is not certain if these are outlying members of the Virgo Cluster to the northwest, or if they are separate islands of stars in uncluttered intergalactic space. The presence of the Milky Way to the east may mean that some of these objects are dimmer than they would be otherwise due to obscuration from the dust and gas of our own galaxy.

Lyra
(The Welsh Harp)

This outstanding summer group is one of the most easily recognizable constellations, hanging like a jewel close to the Milky Way, a companion of Cygnus and of equal interest. Lyra – the Lyre – is a group that celebrates the invention of this musical instrument, or as classical legend would have it, represents the lyre of the famed musician Orpheus. His music was endowed with such power and beauty that he could enchant man and beast with his playing.

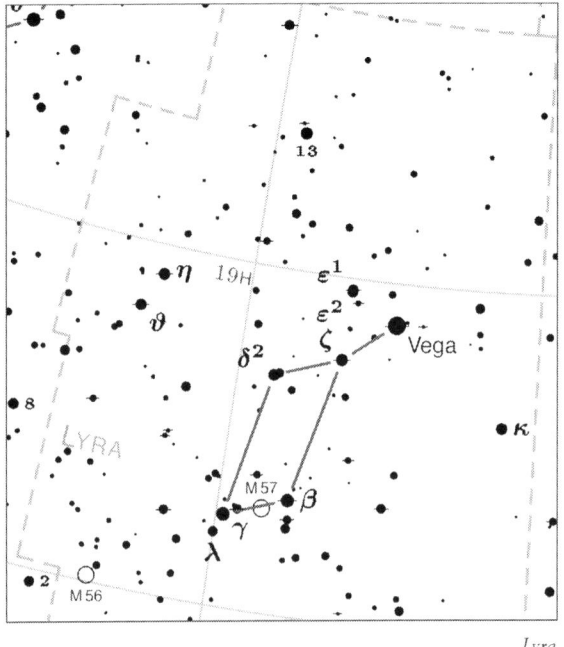

Lyra

In Welsh legend the group was known as Telyn Idris after the giant Idris, an astronomer, poet and king who turned the mountain Cadair Idris into a chair so he could watch the stars and play tunes. It was later associated with the peace of Camelot under King Arthur and the harp has been a favourite instrument of Celts in both Wales and Ireland for its melodious and peaceful sounds. It became known as Telyn Arthur and ever since pubs and hotels have been known as the Harp in honour of this association.

From Snowdonia comes a legend that ties harp, a lake and the sky together. The tale involves Lake Bala, Wales' largest natural lake, and as in many lake traditions, where the lake now lies was once a town which was full of odd, selfish people and ruled by a typically wicked man, who held a party in his palace to celebrate the birth of his first child. A local harper (did you know that a female harp player is known as a harpist whilst a male harp player is a harper?) was ordered to provide entertainment at the party. Although he was reluctant to do so, the harper knew it would be very dangerous to refuse, and he agreed to play for the party goers. As the party progressed the harper heard a strange noise behind him. He turned to see a bird which kept repeating the same word: "Vengeance! Vengeance!" whilst at the same time signalling to the harper to follow it. Withdrawing from the guests, he left the palace and followed the bird up to the top of a hill, where magically he was overcome with sleep and collapsed on the green grass. When he awoke the next morning, he was surprised to find that the town had disappeared and in its place was large lake. As he looked closer he could see that the only thing that had been saved from the town was his harp floating on the waves....

A tale from mid-Wales recounts how a man named Morgan played a harp and though he thought that he made melodious sounds, his neighbours could not stand the tunes which were alternately flat or screeching. One day a bard passing Morgan's house made fun of his playing and singing, and Morgan was distraught. The bard moved on, but that night Morgan was visited by three travellers who strangers to him; they were from the fairy world of the Tylwyth Teg. Although his playing and singing voice were bad, Morgan was a generous man and gave of his own food and drink and bed for the strangers. They repaid his kindness by giving him a magical golden harp and his songs became melodious. In the days that followed many turned up to dance and sing with Morgan, including the bard who had made fun of him. But Morgan used the magic of the harp to harm the bard, breaking his legs by making him dance wildly. As Morgan finished his revenge on the bard, the harp disappeared, taken back by the fairy folk because its music had been used to cause harm. Morgan's harp was placed in the sky out of reach of mortals so its magic could not be used for ill again.

*　*　*

Lyra is another of those constellations where the outline recognisably depicts the object it is meant to represent, and it is easily visible from any part of the Earth's surface. In the late summer night it hangs right at the zenith, with the wonderful blue-white star Vega sparkling in the languid sky. Vega is one of the closest stars to our solar system, lying 26 light years away. It is an A type star shining with 60 times the luminosity of our Sun and has twice its diameter. This is one of few stars which has a circumstellar band of dust, which astronomers theorise may be collapsing to form some kind of solar system.

Close to Vega is an exquisite double star, ε Lyrae, commonly called the 'double, double'. The two main components are easily separated in binoculars, whilst a small telescope will resolve the two main stars into a further two binary systems. This is an incredible sight in such an instrument; all the stars are a wonderful electric blue in colour, despite being catalogued as A type white stars. The close companions look like little sparks of light in a system that will take high powers very well. ε Lyrae is approximately 162 light years away.

The beautiful variable star β Lyrae is one that can be well observed with binoculars. It is an eclipsing variable, much like Algol in Perseus, except that in this instance the components are almost touching one another and are surrounded by streamers of hot gas thrown out by perturbations in the system. The brightness range is not large however, and could be described as a good test of your visual discernment, as the system fluctuates between magnitude 3.8 and 4.1 in a period of 12.9 days. β Lyrae is a giant star lying 960 light years away.

Scan the field between the two bottommost stars of Lyra with binoculars, and halfway along a line drawn between the two you may just see a faint round reddish ring of light, looking like an out of focus star. This is revealed by a small telescope to be the outstanding planetary nebulae M57, the Ring Nebulae, one of the showpiece objects of the summer sky. The only other planetary that can really compare with it is M27 in Vulpecula, but through a small telescope, M57 is a captivating sight. It appears as a reddish-purple ring of seventh magnitude and the central 'hole' is clearly visible as a dark inner ring within this disarming little object. The central star has a magnitude of 14, so is not visible in modest amateur equipment, but the ring of gas is unmistakeable, unlike any other planetary in the heavens.

Messier 57

 This bubble of gas was formed when a Sun-like star puffed off its outer envelope as it reached a point of dying after the onset of Helium burning. Exactly how the envelope is projected into space is a minor mystery, but this ejected shell is visible due to the huge amount of ultraviolet radiation pouring out of the white dwarf central star, causing the gas to fluoresce. M57 lies at a distance of 2300 light years, making the planetary disc roughly half a light year across. This gaseous bubble is an ethereal object and will probably disperse within 20,000 years, which on an astronomical timescale is very short indeed. We are therefore privileged to witness this wonder of stellar nature.

 Within the rich star fields of the Milky Way in the east of the constellation can be found a beautiful globular cluster discovered by Messier in 1779 and designed as number 56 in his catalogue. It is an easy object in binoculars, resembling a fuzz of light shining at eighth magnitude amid a splendid field of stars that almost drown it out. The view through a small telescope is somewhat better, with several outlying stars resolvable, whilst the centre blazes with a white impenetrable light. M56 is 32,000 light years distant and has a collective mass of 90,000 Suns.

 If you are the owner of a pair of giant binoculars or a fairly large telescope, scan the area to the east of the double-double, where, about 7 degrees away, you may be fortunate enough to spot a very rich open cluster known as

NGC 6791. This cluster is a collection of over 500 stars, most of which are fainter than 11th magnitude, but the cluster has an integrated magnitude of nine, which one would expect to be visible even in a modest telescope. However, the cluster is fairly dispersed, and so it has a low surface brightness that usually puts it out of range of small instruments. NGC 6791 resembles that other wonderful cluster NGC 7789 in Cassiopeia for both form and opulence. Once found it is an unforgettable sight.

Lyra contains little else to the casual observer, although those with large telescopes may be able to examine the 13th magnitude E0 type galaxy NGC 6705 in the north of the constellation. For binocular observers, it is worthwhile browsing through the rich star fields of the Milky Way on the borders of Lyra and Cygnus, enjoying the sight of fabulous trains of stars and streams in this attractive area of our summer sky.

Ophiuchus
(Non of the Stream and mother of Dewi Sant)

The constellation of Ophiuchus is a straggling, sprawling collection of second and third magnitude stars that straddles the border of the spring and summer groups. The constellation is one that features in some of the greatest of Welsh stories, especially that of our patron saint, St David.

The constellation is meant to represent Rhiain Non, 'Non of the streamlet' or brook as the circular pattern of the stars of the group herald a pregnant female figure whilst her upheld arms bear the waters of the stream. Non became the mother of this saint after she was raped by the god (or possibly a local chieftain) Sant and after wandering around the Preseli hills she followed a star that brought her to a stone circle where flowers magically grew up around her and a spring opened at her feet, later to become a healing well to pilgrims. The appointed time came and she gave birth to Dewi (or David) during a thunderstorm that poured water over her in torrents, just like the flowing stream across the arms of the heavenly figure.

* * *

Ophiuchus contains a higher number of globular clusters than any other constellation in the sky, which is fortunate, as this motley group of stars holds very little else of interest. The collection of clusters is due to a line of sight effect, as Ophiuchus lies close to the nucleus of our galaxy, so we are seeing clusters orbiting the central bulge of the galaxy, and of course, leaving for their usual domains in the galactic halo. In a small telescope no fewer than 18 of these globular clusters can be seen on any one night, although star hopping to them can be a problem in this constellation of meagre starry guides. A telescope with GOTO facilities is best.

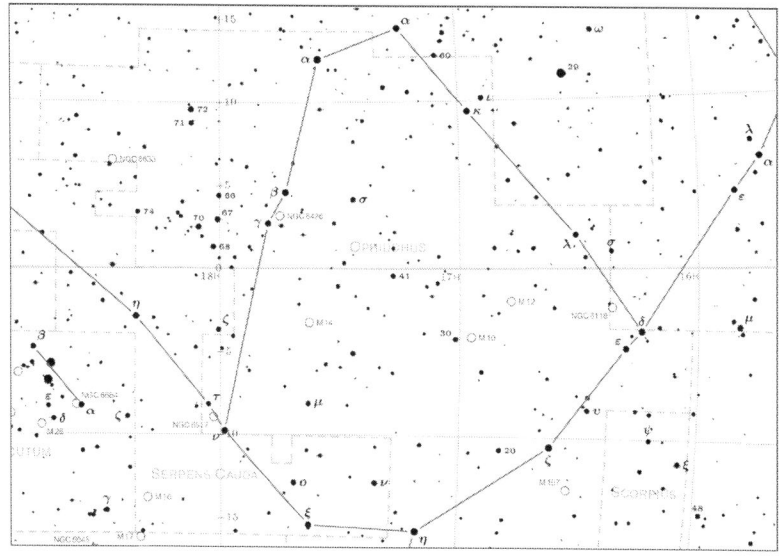

Ophiuchus

One of the most renowned stars in the whole heavens resides within the borders of this constellation, the object known as Barnard's Star. This is a faint red dwarf of magnitude 9.5, and is famous due its large 'proper motion', or the orbital path stars take through our galactic home. This large proper motion is explained by the fact that it is one of the closest stars to us, lying only 6 light years away, and may be a likely candidate to embrace a small planetary system, as the motion of the star shows a slight wobbling, though whether this is due to large Jupiter type planets or a faint companion star is not yet known. Barnard's Star can be found in the north east of the constellation; a good star atlas will give its position.

Turning to the deep sky objects within Ophiuchus, the stunning pair of globular clusters M10 and M12 can be found with a pair of binoculars in the west of the constellation, where in a wide field, they can be seen together. M10 is a compressed globular of 7th magnitude, which through a telescope shows a hint of some resolution into stars. It lies about 14,000 light years away and probably contains upward of 100,000 stars.

Messier 10

M12 on the other hand is a rather sparse globular that can be well resolved in a small telescope with a medium power eyepiece, but to a binocular viewer, is a slightly dimmer, rather hazy object compared to M10. M12 is reputed to lie at a similar distance to M10, although the loose arrangement of this cluster

Messier 12

can make such measurement difficult. Both clusters can be found by taking a line from Lambda Ophiuchi to the star 30 Ophiuchi, where M12 lies midway along, and M10 lies close to 30 Ophiuchi.

The globular cluster M9 is a small round patch of nebulous light lying close to the star η Ophiuchi, and glowing at magnitude nine. It is one of the smallest globular clusters in the sky in terms of diameter; apparently it is some 60 light years across and is 24,000 light years distant. Another Messier cluster is M14, a globular cluster of 9th magnitude, which can be observed as a nebulous patch in binoculars. The view through a telescope is not much better, as the stars all appear unresolved, it simply being a white round spot against the darkness of space. M14 is fairly remote at 30,000 light years.

Within Ophiuchus are two galactic clusters that can be observed well with a telescope or binoculars. The first is NGC 6633, RA 18h 27m 42s Dec 06°34m; a collection of 65 stars in a bright compressed field with an integrated magnitude of 5, thus making it a naked eye object, though it has never been reported as such. It is slightly elongated in form and its stars are mostly A type giants glowing softly at 7th magnitude.

A rather scant group well worth searching for with binoculars is the small cluster I.C. 4665, a group of 20 stars in an area twice the size of the full moon, and lying in an obscured area of the Milky Way. Most of the stars are of 7th and 8th magnitude, so it should not be difficult to see, although a telescope will tend to look through the cluster rather than at it.

IC 4665

Ophiuchus also encloses a bright planetary nebula; but this is a rather difficult object to see with either binoculars or a small telescope due its very small diameter, even though it shines at magnitude nine. This is NGC 6572, at RA 18h 12m 06s Dec 06°51m, an out of focus greenish star in the star fields of the Milky Way to the north east of β Ophiuchi.

Resuming our search for globular clusters, we come across the gorgeous M19, a 7th magnitude bundle of stars that is a pleasure to see in a pair of binoculars or a telescope. Some small outlying stars can be resolved with a telescope, but as M19 is located in a very rich part of the Milky Way it is uncertain whether these stars are associated with the cluster or are merely foreground objects. M19 lies 29,000 light years away from our Earth.

There are several more globular clusters worthy of scrutiny in Ophiuchus, although recourse to a good star atlas is recommended to find them. Close to the Ophiuchus, Sagittarius and Scorpius borders is the most intriguing area of dust clouds in the whole of the Summer Milky Way. If you centre your binoculars on θ Ophiuchi, then slowly sweep the area; you will come across some of the most wonderful dark nebulae in the heavens, which are shown to advantage in the wide field of binoculars rather than the narrow view of a telescope. Two of these objects are B 78 (Barnard's catalogue of dark nebulae), commonly called the Pipe Nebulae due to its resemblance to a smoker's pipe, and the incredible B 72, the S Nebulae, looking like a black letter S against the background stars. Both are faintly visible as obscuring masses against the brightness of the Milky Way.

Sagittarius
(Gwrhyr, Daghda and Taranis)

Sagittarius is without doubt one of the greatest, if not the greatest constellation in the whole sky. Its prominent position sitting over the brilliant star fields of the nucleus of our galaxy ensure that it is very rich with deep sky objects that are visible in binoculars or small telescopes, whilst the glory of the Milky Way is evident even to the naked eye.

Some sources state that this constellation takes its place in Welsh mythology as that of Gwrhyr, the interpreter between man and animals in the tale of Culhwch and Olwen from the *Mabinogion*. As animals were meant to have magical powers and become the tutors of those wise enough to watch them, so Gwrhyr was wise enough to give counsel to the band of seekers in the story so that they may find Mabon, the man who could control the hunting dog

Drudwyn. He did so by changing his shape through a pantheon of fantastic animals until at last Mabon appeared and offered to help Culhwch in his quest. In some sources it is suggested that the constellation is Mabon who can take the likeness of a stag and use his godlike strength in various ways.

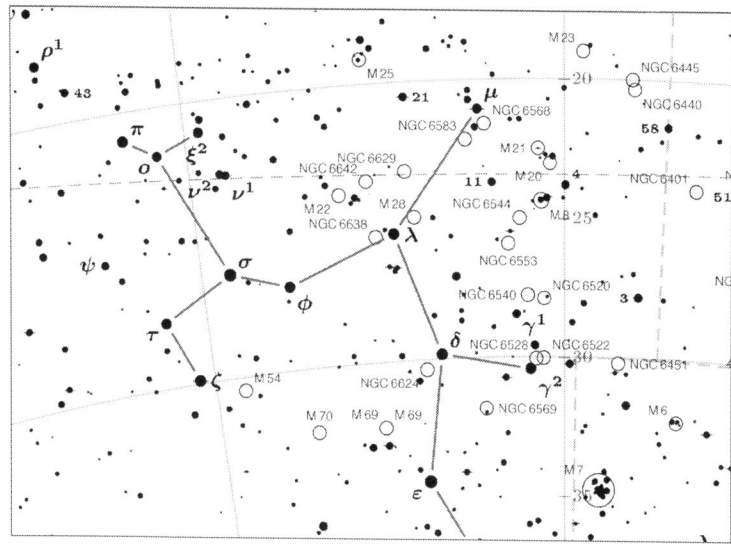

Sagittarius

In all probability the constellation is older than the Arthurian story. Across the Celtic world, stories of shape shifters and gods who can change their appearance at a whim are interwoven in Welsh, Irish and Gallic tales. The Irish Daghda or the Welsh Taranis are both seen as the same god, one who is responsible for storms, thunder and lightning. Both are the equivalents of Jupiter, or Zeus, who was also a shape shifter, changing to complete one amorous conquest after another. The fact that both Taranis and Daghda were thought responsible for the largest planet in the solar system and at the same time the gods responsible for the Sun are revealed in Sagittarius as it is within this constellation that the Sun reaches its lowest point on the horizon before being reborn and ascending after the winter solstice.

This power to recreate the Sun and the new cycle of the year and its immortal attributes sits well with the 'good god' of the high heavens. Some sources allude to the Sagittarius group as Taranis, as the December Sun, rising from this group promised life, fertility and prosperity and thus the constellation held a special prominence. Many barrows and dolmens are aligned to the rising of the winter solstice Sun and no doubt religious rites played a part in celebrating the special position of the constellation to the ancient Celts.

The figure of Sagittarius is supposed to delineate a Centaur, that half man, half horse creature of Greek mythology, apparently holding a bow and an arrow, drawing it back and ready to let fly at anything within range. To the ancients, this was the centaur Chiron, who taught Hercules all his skills, only to be accidentally killed at his hand by a poisoned arrow. The centaurs were immortal beings, but the agony of the poison was too great to bear so Chiron begged Zeus to release him from his immortality. Zeus obliged, and at the request of his athletic son, placed the centaur in the sky in memory of his mentor.

* * *

To the observer, the main body of Sagittarius is composed of eight stars that make up the asterism known as the Teapot, as it resembles this humble object more than it does a centaur! The Teapot is the key to unlocking the deep sky secrets of the group, as most of the objects lie along the Sagittarius Milky Way like steam rising from its spout. All manner of objects can be found in this sensational constellation, ranging from star clusters, globular clusters to bright nebulae, as well as the incomparable grandeur of the Milky Way star clouds.

The brightest nebula in Sagittarius is the inimitable M8, the summer equivalent of M42 in Orion. M8 is commonly known as the Lagoon Nebula, and is an easy object in binoculars, a patch of glowing gas elongated along the east-

Messier 8

west axis and containing the beautiful star cluster NGC 6530, a group of over 60 bright new O and B type stars that have formed from the nebulae. Through a telescope, M8 glows eerily as a fifth magnitude swirl of blue-white light that seems to radiate out of a low power eyepiece, whilst larger telescopes will show the famous dust lane bisecting the nebulae and creating the lagoon that prompted Herschel to name it so.

Just 2.5 degrees north of M8 lies another nebulae of fame, M20 the Trifid Nebulae, so called due to the fact that its three dust lanes appear to bisect it like a flower. M20 is not an easy object for a pair of binoculars, but a small telescope will suffice to show it well. The centre of M20 is well marked by what appears in a telescope to be a single star, lying on the interior portion of one of the dust lanes, although high powers are needed to see this apparition. Both the Lagoon and the Trifid Nebulae are tremendous distant from us, over 4100 light years for M8 and over 5200 light years for M20. If they lay at equivalent distance, M20 would be the equal of M8 in terms of brightness and area.

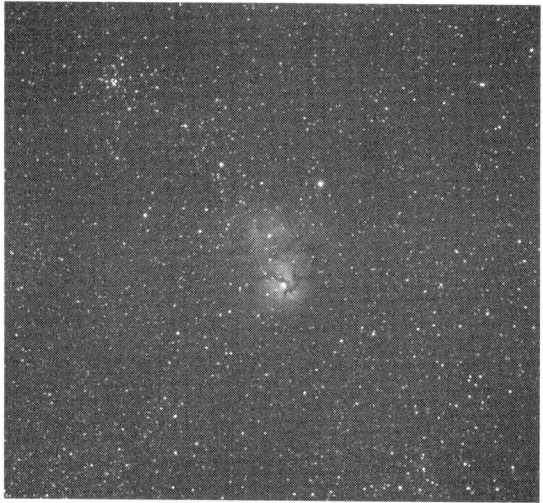

Messier 20

Just above the 'lid' of the teapot is one of the most detailed globular clusters available to owners of small telescopes. This is M22, a globular that is currently diving through the disc of our galaxy as it orbits the galactic centre. M22 glows at 5th magnitude and can be seen as a woolly ball of light in a pair of binoculars, whereas a small telescope will reveal an incredible ball of stars, the outer members of which are easily resolved. The globular is apparently very close by, only 10,000 light years away in a very dusty part of our galaxy. M22 contains over 175,000 stars including a high number of cluster variables. To the west can be found another globular cluster mentioned in the Messier catalogue, M28. Although this is not as magnificent as its near neighbour, it is nevertheless an interesting object, glowing at eighth magnitude in a very rich portion of the Milky Way.

Messier 22

Above these two objects is one of the greatest naked eye star clouds in the galaxy. This is M24, or the small Sagittarius star cloud, a rectangular aggregation of over a million stars, beaming softly its radiant light. It is very easy to capture in a photograph, and binocular exploration of this amazing cloud will uncover hundreds of starry points in a three dimensional aspect that is truly delightful. The cloud contains a star cluster at its northern border, NGC 6603, which has a magnitude of 6.5, and contains over 60 stars, though it is not definitely known if this object is actually a member of this star cloud or if it is a foreground object. M24 is, as one would expect, extremely distant, perhaps in excess of 10,000 light years, and measures 560 light years across its major axis.

To the west of this star cloud is an exquisite star cluster known as M23, which embraces over 100 bright stars in a rich scattered group. Most of the stars are around ninth magnitude, but the cluster can be seen as a hazy patch against the background stars with a pair of binoculars. A telescope will reveal its full glory, with radiating streamers of B type stars that are a pleasure to witness. M23 is approximately 2000 light years away in this heavily obscured

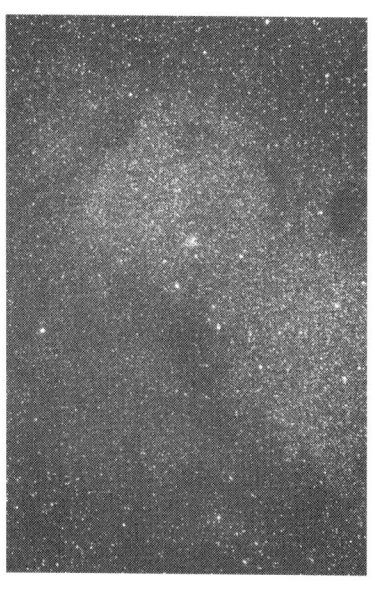
Messier 24

region of space. To the east of the M24 star cloud is another Messier cluster, M25, which is a much sparser group than M23, though it holds about 50 stars in a rather pleasing field. This would be a showpiece object were it not for the incredible array of other deep sky wonders to be found in Sagittarius. M25 is also about 2000 light years away from our solar system.

If the observer travels northwards of this trio of objects, you will encounter another bright nebula that is a stunning sight in binoculars or a small telescope. This is M17, the Omega Nebulae, so called by William Herschel as it reminded him of the Greek letter omega. In a small telescope its alternative name of the Swan Nebulae is far more appropriate, as it does look like a swan gliding along the river of stars that is the background to this extraordinary nebula. M17 shines at sixth magnitude, and is a large cloud of ionised hydrogen gas at a distance of 5000 light years, giving the nebulae a diameter of about 40 light years along the major axis. Interestingly, M17 contains no star cluster; presumably the stars responsible for its illumination are well hidden inside the gaseous envelope of their birth.

As I have said, Sagittarius has several globular clusters of note, many of which are Messier objects that are an easy target for binocular observers. Such a cluster is M54, an eighth magnitude object close to ζ Sagittari, although not much detail can be seen in either binoculars or a small telescope, some outliers

Messier 17

may be spotted during periods of steady seeing. Its chief delight is that it is probably the most distant of the globular clusters observed by Messier, at over 87,000 light years away. Another globular of note is M55, a rather loose looking cluster some 7 degrees east of M54. In a small telescope the cluster shows some resolution into stars, and even binoculars will show it as a rather lively object of seventh magnitude. M55 is about 18,000 light years away.

One Messier object that should not be missed is M75, a globular cluster in a blank area of sky in eastern Sagittarius. This cluster is not particularly remarkable; it cannot be seen to advantage in binoculars, and is almost irresolvable in a small telescope. Messier 75 is 67,000 light years from us. This little patch of nebulous light of 9th magnitude hides a wonderful collection of about 500,000 stars, which would make it a stunning object if it lay as close to us as M13 or M22. Binocular observers will find it easier to track down, as their wide field of view shows such objects to advantage over the small field of view of the average telescope.

Sagittarius contains many galactic or open clusters that can be seen in a small telescope. One of the loveliest is NGC 6645, which is to be found just to the north of M25. NGC 6645 is a rich compressed cluster of 75 stars of the 11th magnitude that radiate at a combined magnitude of 9. NGC 6568 is another fairly rich cluster of some thirty stars of 11th magnitude, which can be found near the star μ Saggitari. One staggering cluster is NGC 6520, a sparkling field of 25 O and B type stars in a very rich portion of the Milky

Way, making it rather difficult to separate from the background stars.

All observers are encouraged to enjoy the delights of this, the richest portion of the Milky Way. Scanning the area with binoculars is very rewarding as they give an impression of immensity that no telescope can convey. Dark dust lanes and nebulae that blot out the light of the stars are best examined with binoculars, and such objects grow in profusion around the nucleus of our galactic residence. The bright star clouds of the Milky Way rise to a crescendo near the star γ Sagittari, where we are looking almost directly into the core of our galaxy. It is a pity that the actual centre is so obscured as to be invisible to human eyes, but the glory of the Sagittarius star clouds makes up for this lack.

Through binoculars, starry clouds tumble out of the eyepiece, starry groups swarm along the river of stars that mark the arms of our galaxy, whilst here and there can be found beautiful nebulae that inspire and reward the patience of the astronomer. This unbelievably rich area can easily be arrested photographically; even relatively short exposures will enable clouds like M24 and nebulae such as M8 to be recorded, to be admired at a later date.

Sagitta
(The Arrow)

The constellation of Sagitta is a tiny grouping of five fairly bright stars that form the outline of the Arrow this group is meant to represent. The constellation was introduced by the Greeks, but is so conspicuous that it is hard to see how it was not noted by other more ancient civilizations. It is said to represent one of the arrows Hercules used to kill the Stymphylian birds as part of his twelve labours; or alternatively, the arrow of cupid.

The arrow has several meanings to the ancient Welsh; in some tales it is the figure of the arrow that hunters use to bring down the Greylag goose. Many ancient arrows were fitted with goose feathers and Michael Bayley recounts in *Caer Sidhe* how the constellation of Cygnus is known as the Greylag goose. Perhaps an old Welsh archer has fired at the bird and only just missed the target. In the *Annals of Wales* Gerald of Wales writes of the exhumation of the body of Harold Godwinson, the last Saxon king of Britain and notes the arrow that felled him, though his body was as fresh as the day he was killed.

It also recalls one of the many tales of King Arthur and is said to be the arrow that killed him. Arthur's last battle apparently took place near Snowdon and ended with his disappearance into the mist wreathing the mountain after being mortally wounded by an arrow at the Bwlch y Saethau or the Pass of

Arrows. How the arrow got into the heavens is a mystery, though some legends recount Morgana or Merlin removing it and throwing it into the sky. Arthur is said to have been buried where an arrow he shot during his dying moments landed, and some stories recount that the arrow never landed at all but went into the sky in symbolism of Arthur's soul ascending to the heavens. Whether these stories have a part in Arthurian legends is difficult to decide; there may be some confusion with Sir Bedivere throwing Arthur's magical sword Excalibur into Llyn Lladaw, one of the lakes near the mountain.

* * *

Sagitta lies along the Milky Way between the rich constellations of Aquila and Vulpecula, and contains two beautiful star clusters and two very unusual stars that will interest the casual observer.

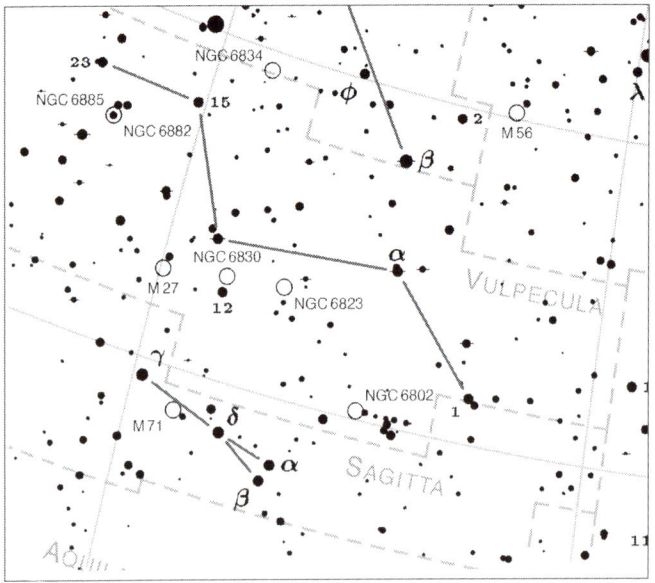

Sagitta

The first of the star clusters is the incredible compact group of stars known as M71, which can be found along the shaft of the arrow. In binoculars it is a bright hazy spot of blue-white light, almost merging at the edges with the Milky Way, but its full glory is revealed by a small telescope as a blaze of seventh and eighth magnitude stars, so as to look like a loose globular cluster. It most resembles M12 in Ophiuchus. The opulence of the cluster has led to arguments over its classification as to whether it is a sparse globular or an open cluster. The matter was finally settled when the spectra of the cluster was taken, showing its members to be metal-deficient population II objects. M71 is a very distant open cluster lying 13,000 light years away, and containing upwards of 400 stars.

Messier 71

In the same low power field can be seen a rather unconfined gathering of eighth magnitude stars known as the cluster H20. It is a loose aggregation of 25 stars, and can be well seen in giant binoculars, where the wide field tends to draw this object together, whereas the telescope looks 'through' the cluster.

Just to the south of the star θ Sagittae can be found an extraordinary type of variable star. This is the object FG Sagittae, a ninth magnitude star that has risen from thirteenth magnitude over the last few decades. It is associated with a small nebula, which is invisible to amateur instruments, but this nebula appears to give evidence that FG Sagittae is emerging as a new, young object in its own right, in almost the same way as the star P Cygni. FG Sagitta is over 3000 light years away in a crowded field of the Milky Way.

The other star of note within this constellation is WZ Sagittae, an excellent example of a recurrent nova, very similar to T. Coronae Borealis. Although the period of outburst is irregular, they seem to occur every 32 or 33 years, the last such blow-up occurred in 1978. These outbursts have not yet reached naked eye magnitude however, although during the last three occasions the star has flared up to 7th magnitude, and so will become a 'binocular nova' again. WZ Sagittae is invisible to most amateur equipment as its minimum magnitude is 16, but getting to the know the star fields around the object will enable the sharp-eyed observer to know immediately if there are any interlopers.

Scorpius
(The Addanc)

Scorpions are not common to Wales and as the full extent of the classical constellation does not rise in this country, the stars symbolized something else. In some tales the creature is said to be the Llamhigyn y Dwr, or the Water Leaper, an evil creature that lived in swamps and ponds. Apparently it was a sort of giant frog with a bat's wings instead of forelegs, no hind legs, and a long, lizard-like tail with a sting at the end. It jumps across the water using its wings, and takes people unaware of its presence in ponds or mountain tarns.

In a similar fashion, the Afanc, or Addanc, was said to be a monstrous creature that preyed upon anyone who was foolish enough to fall into or swim in its lake. One of the earliest descriptions of the Addanc is given by Lewys Glyn Cothi, who described it as living in Llyn Syfaddon, now known as Llangorse Lake. In many other tales it lives in Llyn Cwm Ffynnon in north Wales, but it seems the monster can truly be called a common Welsh motif.

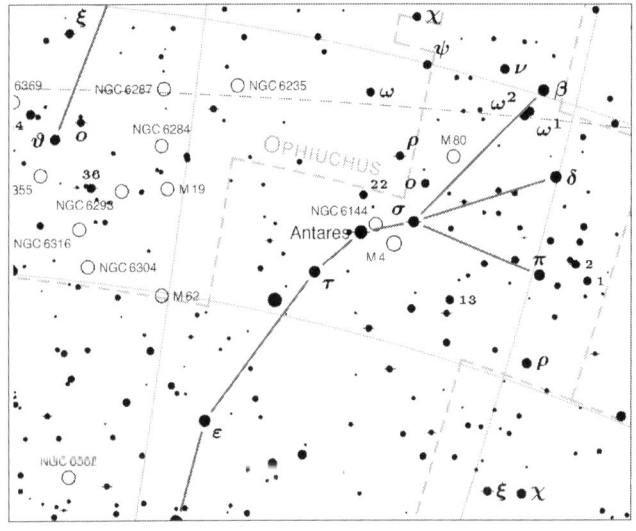

Scorpius

The legendary Hu Gadarn rescues villages from flooding by tricking the monster, rendering it helpless by having a maiden sing it to sleep in her lap. Whilst the creature slept, Hu and the villagers bound the animal in chains. The Addanc was awakened and like any trapped animal went berserk and crushed the poor young woman in its thrashings. Nevertheless, the Addanc was secured and the team of oxen under Hu's control led it away to Llyn Cwm Ffynnon, or the lake of the well, perhaps a reference to the fact that it could not escape, as if caught at the bottom of a well.

In the *Mabinogion,* the Addanc of the Lake resided in a cave near the Palace of the Sons of the King of the Tortures. The palace is so named because the Addanc slays the three sons of the king each day, only for them to be resurrected by the maidens of the court. In the tale, Peredur son of Efrawg (Sir Percival in Arthurian legend), asked to ride with the three chieftains, who sought out the Addanc daily, but they were afraid that innocent blood would be on their hands as if he was slain they would not be able to bring him back to life.

Peredur continued to the cave alone wishing to kill the creature to increase his fame and honour. On his journey he met a maiden who told him that the Addanc would slay Peredur through cunning, as the beast is invisible and kills his victims with poison darts. However, the woman gave Peredur a magical stone that made the creature visible. Peredur ventured into the cave and with the aid of the stone, found the Addanc and ran it through with his lance before beheading it. The Addanc then took its place as a mystical creature in the sky.

* * *

The constellation lies along the southern, wealthy portion of the Milky Way, the centre of our galaxy, being just over the border in Sagittarius. Scorpius contains quite a few objects of particular interest to those equipped with modest instruments such as binoculars or a small telescope, and is a home to excellent star clusters.

The first, astounding, object in Scorpius is α Scorpii, the red giant star Antares. The name means 'Rival of Mars', as the star's ruddy hue is very like that of the warrior planet. This is one of the greatest stars in the sky, spectral type M1Ia 900 times the diameter of our Sun, with a mass of 12 times the solar mass, and a luminosity of 60,000 times that of the Sun. So large is Antares, that, if it were placed in the position of our Sun, then the Earth would be orbiting within its atmosphere! Antares has a green companion, thought it is not an object for average equipment. This beautiful first magnitude star is a beacon of the summer sky and lies at a distance of 550 light years.

Not far from Antares is a spectacular globular cluster for either binoculars or a telescope. This is the awesome M4, a globular cluster that is simple to find, looking like a smattering of luminous points of light in a low power eyepiece. This object and M22 in Sagittarius are the most easily resolvable globulars in the sky as seen from Britain. M4 glows brightly at magnitude 6 and comprises of upwards of 50,000 stars, making it rather sparse by average globular standards. It is presently crossing through the disc of our galaxy, and lies at a distance of 7200 light years, although this distance is open to question due to the heavy obscuration in this area.

Messier 4

To the northwest of M4 can be found another lovely globular cluster, M80. This is not as bright or as well resolved as M4 as it lies considerably further away, over 32,000 light years in fact. It glows as a misty round patch of eighth magnitude light and can be seen in either a pair of binoculars or a small telescope, though detail is lacking.

Scan the heavens to the north of Antares with binoculars and you will find that there are fewer stars in this region than in any other part of the Milky Way. This is due to the presence of a beautiful nebula centred on the star ρ Ophiuchi just over the border of Scorpius. This nebula is Herschel's distinguished 'Hole in the Heavens', and is a vast cloud of dust and gas, obscuring the background stars of our galaxy. The nebula has wonderful colours of all descriptions, but unfortunately, they only be captured by long exposure photography.

In the eastern part of the constellation, above the tail of the scorpion, can be found two of the most absorbing star clusters in the whole sky. Despite their low elevation, at culmination and with a little luck they are visible in binoculars or a small telescope where they reveal themselves to be beautiful showpiece objects. These are the star clusters M6 and M7, lying in a particularly rich part of the Milky Way that extends into the great Sagittarius star cloud.

M6 is the more westerly cluster of the two, and also the more magical, shining lustrously at magnitude 5.5, which would suggest that it could be a barely discernable naked eye object from southern climes. M6 holds over seventy O and B type stars in a beautiful configuration that has earned it the nickname the 'Butterfly cluster', as the outline of this insect is immediately evident, even in a pair of good binoculars. Through a telescope, the view is staggering, starry points radiate out of the eyepiece, the centre of the cluster is bright with stars, although the butterfly shape is slightly lost if one uses too high a power. M6 was discovered by P.L. de Cheseaux in 1746, but it appears to have been known from ancient times, reinforcing the view that it is a naked eye object, but is too low to see in this way from Britain. M6 lies in the Milky Way at a distance of 1600 light years.

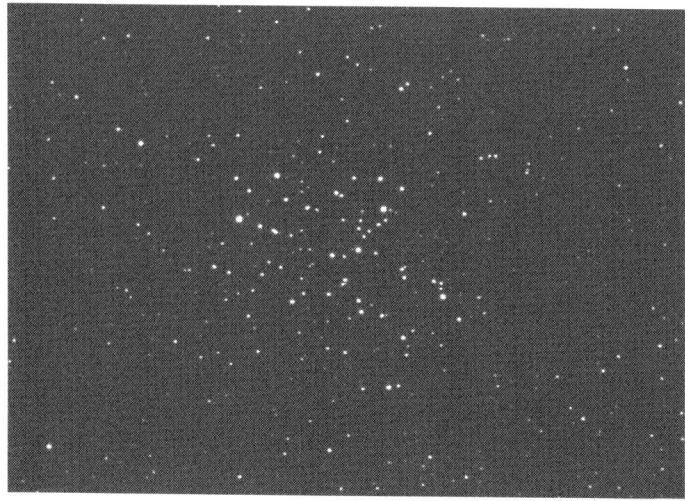

Messier 6

Its companion cluster in this wonderful area of sky is the brilliant and compact group designated M7. Again this cluster is visible to the naked eye, having been noticed by eminent astronomers of antiquity. M7 lies in a very rich field of Milky Way stars that provide a perfect backdrop for this jewel-like cluster. It contains over eighty stars in a compressed group, which is easily visible in binoculars, and like M6 is stunning when seen through a telescope. It lies much closer to us than its bright neighbour M6, approximately 980 light years distant. All its stars are hot young B or A type stellar giants. An interesting juxtaposition of objects can be found in this region with a fair sized telescope within the same low power field as M7. This is the globular cluster NGC 6453, an 11th magnitude nebulous patch of light that is difficult to discern against the sweep of Milky Way stars in this beautiful region.

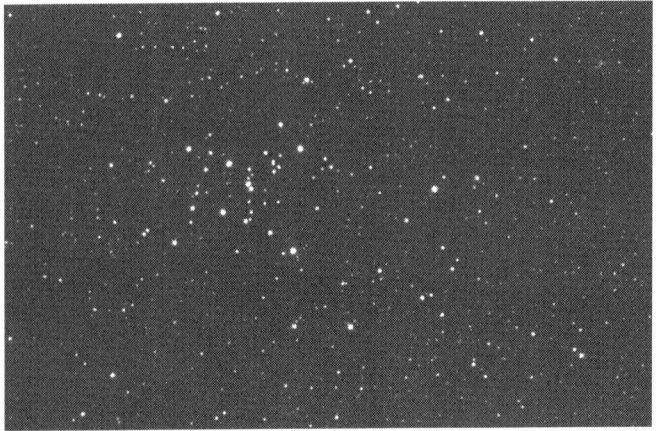

Messier 7

Another globular worth locating is the cluster designated M62, an eighth magnitude blob of light that shows a little resolution into stars in a small telescope. M62 lies about 22,000 light years away and probably contains upward of 400,000 stars. A star cluster to investigate is NGC 6451, RA 17h 50m 42s Dec -30°13m; an interesting group of fifty faint stars that shine with a combined luminosity of 8th magnitude in a rich compressed group low on the horizon.

Scutum
(The Shield of Arthur)

Scutum is one of the smallest constellations in the northern sky, yet it is packed with all manner of deep sky objects guaranteed to thrill and captivate the observer. The constellation was created by Hevelius and is supposed to represent the shield of the Polish king Johann Sobieski, who made a spirited defence of Vienna during the Turkish invasion in the sixteenth century. Unlike most of Hevelius' creations, Scutum is not an obscure blank area of the sky, but one of the brightest parts of the Summer Milky Way, as it is composed of a radiant star cloud that is perceived as the brightest patch of nebulous light along this rivulet of stars.

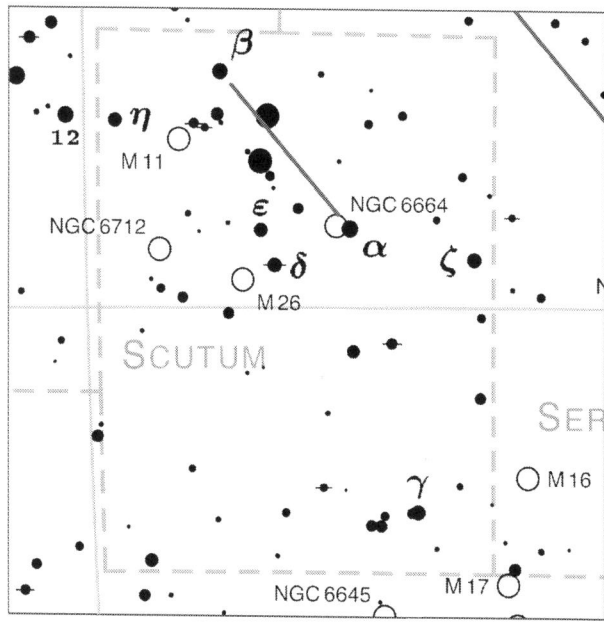

Scutum

In tales from Wales this constellation is associated with the magical shield of King Arthur, known as Wynebgwrthucher (the face of evening). The shield was blue emblazoned with three gold crowns although Geoffrey of Monmouth writes that it featured a likeness of the Virgin Mary. This is confirmed by the historian Nennius who talks of Arthur carrying the likeness of Mary on his shield at the battle of Gunnion Fort. Wynebgwrthucher gave him supernatural power to vanquish his enemies and after Arthur's death the shield disappeared, as did Excalibur, and was placed in the sky.

* * *

The proximity of the Milky Way ensures that the constellation is a delight for observers with binoculars or small telescopes, whist owners of giant binoculars will have a wonderful time sweeping the star clouds of this rich region of the sky.

One of the showpiece objects of the summer sky is to be found close to a knot of bright stars at the northern end of the constellation, amongst which β Scuti is the brightest. Lying in the same low power field is a blaze of starry light that is extremely compressed, looking an almost solid object through binoculars. This is the incomparable M11, the Wild Duck star cluster, so called by the amateur observer Admiral William Smyth (1788-1865), as its V shape reminded him of a group of ducks in flight. M11 is an astounding cluster that probably contains over 800 stars, making it the richest galactic star cluster in the Messier catalogue. It can be resolved in a telescope as an eighth magnitude cluster of stars with a central blaze of half resolved objects in a very crowded region, with a lovely red star just at the apex of the V. The cluster is very remote, about 6200 light years, hence its crammed appearance in the eyepiece.

Messier 11

Just above β Scuti is a famous long period variable star known to observers as R. Scuti. This is an intensely red star that can be followed through its rather irregular period with either binoculars or a small telescope. The duration is not well known but appears to fluctuate between magnitude 6 and 8.5 in roughly 140 to 150 days. It is yellow supergiant of G0 type lying at a distance of 1500 light years and shimmering with a luminosity over 9000 times greater than the Sun.

Travelling south into the heart of the constellation, the observer comes to another bright object lying southeast of the star ε Scuti. This is M26, a compact star cluster of some twenty-five bright stars in a very rich field of the Milky Way. So lavish is this region that it is hard to discern where the cluster ends and the Milky Way begins. It is a fairly remote cluster, apparently lying at a distance of 5000 light years and can be seen with binoculars as a misty patch of light. A small telescope will bring out the best in this lovely cluster, revealing the stars to be bright O and A type giants, shining with luminosities thousands of times that of our Sun.

Messier 26

To the east of M26 can be found a small globular cluster, NGC 6712. This small ball of light can be seen with a properly mounted pair of binoculars as a round conglomeration of stars against the concentrated starlight of the Scutum star cloud. In a small telescope it looks like an unresolved mottled sphere of white light, nicely compressed and separated from the Milky Way. It is uncertain whether NGC 6712 is a foreground object compared with the cloud or if it is actually passing through it. If it is a foreground cluster then its distance must be between 5,000 and 8,000 light years.

Close to the star α Scuti is a cluster of twenty-five 10th magnitude stars in a scattered group known as NGC 6664. α Scuti is not actually a member of this star cluster, it is merely a foreground object that is another one of nature's chance alignments. Binoculars should show this rather large and faint cluster, whilst a small telescope will sharpen it nicely. In the same binocular field may be found another star cluster, the group NGC 6649, a pleasing assembly of some thirty-five stars, most of which are of 11th magnitude, but shining collectively at magnitude 9. Through a telescope the cluster is a lovely rich object with a beautiful 6th magnitude double star to its southwest that is easy to split, although both components do not exhibit much difference in colour, as they are both bluish B type giants.

The whole area of Scutum is a brilliant blend of star clouds and deep sky objects in perfect balance. The observer will be drawn to this area throughout summer as the beacon of this incredible galactic condensation draws the eye. The Scutum star cloud is relatively easy to capture photographically as it glows at fourth magnitude with the collective light of billions of stars. Along the western edge of the cloud is the dark lane of the Cygnus Rift, and where it meets the cloudy fringe of stars, it breaks up into dark streamers and intrusions that are easily visible on a photograph. Some of these dark clouds can be seen if you carefully scan with binoculars along the periphery of the star cloud, where the wealth of stardust suddenly gives way to the empty blackness of the rift.

The distance to the Scutum star cloud is not accurately known, but it must be well in excess of 7000 light years, whereas the Cygnus Rift probably lies between 5000 to 5500 light years away, although these figures can by no means be taken as final. Within the cloud are two bright condensations that have been allocated NGC numbers. These are NGC 6682 at RA 18h 41m 36s Dec -04°46m and NGC 6683 at RA 18h 42m 12s Dec -06°17m. These objects are in all probability not true galactic clusters in the full sense of the word, but are more likely areas of the cloud, perhaps containing a higher than average concentration of bright stars, or simply bright clumps of stars lying on the closest edge of the cloud to us.

Considering the beauty and detail shown in the Scutum star cloud, it is a particularly fruitful place to search for novae; indeed, several novae have been discovered here in the recent past. The confusion of stars in this rich area is one that will take a little patience to learn, but is a worthwhile project.

Serpens
(The Birth of Hu Gadarn)

The lovely group of Serpens can be found directly underneath and to the west of Hercules, where the pointed head of four stars is an indicator to the rest of the constellation. Serpens – The Snake – is actually two constellations, Serpens Caput being the head of the snake and Serpens Cauda the tail. On ancient star maps the whole constellation winds through the stars of Ophiuchus, The Serpent Bearer, depicting the writhing snake that twists and turns in his arms. Originally Serpens was intended to have medical significance but its representation has since been corrupted so that it resembles the dragon of the Draco legend or, as early Christians thought it, the serpent that seduced Eve and led to man's downfall.

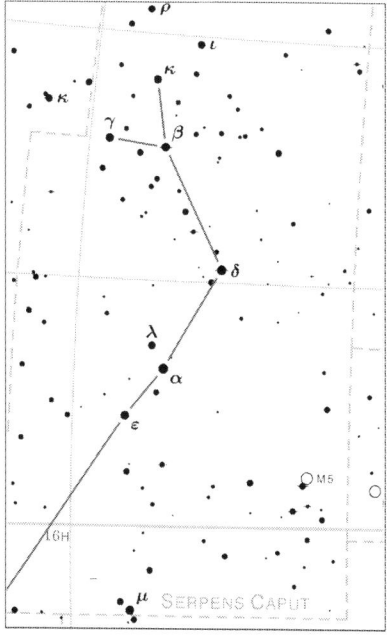

Serpens

Welsh tales recount that the Serpent was the soul of the predecessor of Hu Gadarn, and it guarded the tree which magically gave birth to Hu. He was then nursed by a lady associated with the moon so it appears that this Welsh hero has godlike attributes and roots. The constellation is also associated with Dewi Sant in the tale of Non.

***

Serpens contains few stars that rise above magnitude 2.5, but is nonetheless an interesting constellation due to the presence of two of the greatest showpiece objects of the northern sky, the globular cluster M5, and the 'Eagle Nebulae' M6.

M5 lies in a rather barren field of stars to the southwest of Alpha Serpentis, but it can easily be seen as a sixth magnitude object in binoculars, looking like a condensed blot of starlight. A small telescope will uncover far more detail in this radiant cluster; several of the outlying stars are resolvable, giving M5 a three-dimensional appearance. The globular contains upwards of 500,000 suns and is set at a distance of 24,000 light years. It is usually regarded as one of the oldest globular clusters known, most of its stars having evolved away from the main sequence, especially those with high apparent luminosities. Estimates put M5 between 10 and 13 billion years old, perhaps slightly older than most population 2 stars of the galactic halo.

Messier 5

M16 is a beautiful nebula that contains the star cluster NGC 6611. Unfortunately, the nebula is a little too faint to detect with small telescopes, although giant binoculars may show it as a hazy patch. It can easily be captured on photographic film, where it really does resemble an eagle swooping down on its prey. The cluster, NGC 6611 is easily seen in a small telescope as a compact group of over 100 stars, the brighter members of which are O and B type giants, newly formed from the dust and gas of the nebulae. The nebulae itself is over 70 light years in extent, larger even than M42 in Orion, but is far smaller visually and photographically due to its great distance, some 7000 light years from us. The environment surrounding M16 is full of knots of starlight and dark nebulae, the Milky Way being especially rich in this region.

A beautiful star cluster worth exploring with binoculars is I.C.4756, a very wide cluster twice the diameter of the full moon that is best seen with such an instrument, as a telescope simply looks 'through' the cluster rather than at it. It lies in the Milky Way a few degrees above M16, and is beautifully shown on photographs of the area.

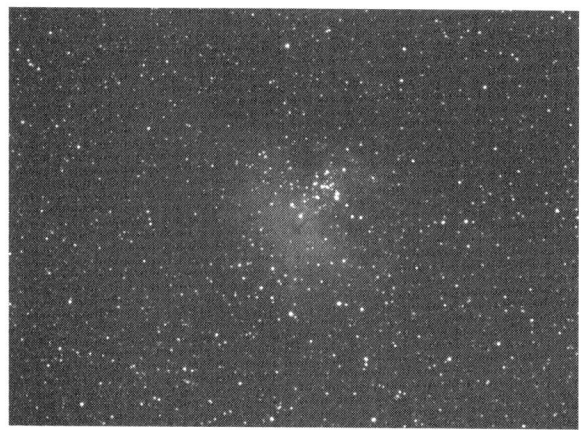

Messier 16

For those with above average telescopes, try looking for two globular clusters, NGC 6535 (RA 18h 03m 48s Dec -00°18m) and NGC 6539, (RA 18h 04m 48s Dec -07°35m) both of which lie in an area of the Milky Way that is heavily obscured by dust. Both clusters shine feebly at 12th magnitude. For those with binoculars, try scanning the rich fields of this part of our galaxy, as this area is close to the galactic centre, and all manner of starry condensations can be glimpsed, delighting the observer with vivid flashes and the mottling of countless suns.

Vulpecula
(The Fox)

This rather dim and obscure constellation was created by Hevelius to fill another blank area of the sky between the constellations Cygnus to the north and Aquila to the south. Its Latin name means 'The Fox', but on older star maps it is called Vulpecula et Anseris, or the 'Fox and the Goose', although how this group represents either one of these creatures is not easily apparent. Ancient Welsh sources refer to the fox as Llwynog, meaning 'bushy' in reference to the tail of the animal though the association with this rather later addition to the skies is sketchy at best.

* * *

Vulpecula is packed with star fields and objects of intense interest even to amateurs with modest instruments. Additionally, binocular observers have discovered many novae, in recent years within this enchanting constellation, which initially may have remained below naked eye magnitude threshold.

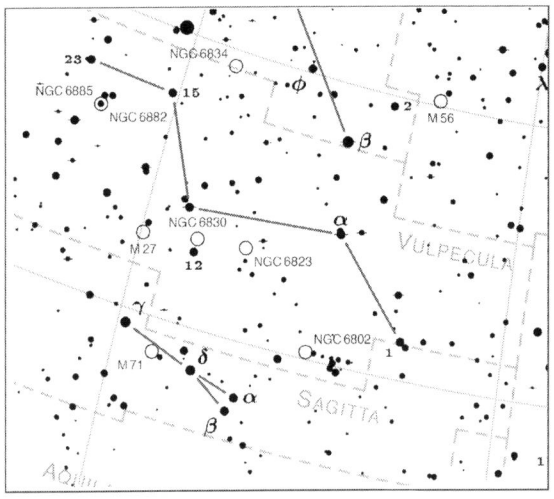

Vulpecula

One of the most beautiful small asterisms in the whole sky can be found with binoculars in the western part of the constellation. If you centre the star Albireo or β Cygni in your field of view, then sweep southwards, you will recognise the brilliant shape of the Coat Hanger as this group of ten stars is called, glistening against the background stars. Its catalogue name is Collinder 399 and is a truly arresting sight, even though it 'hangs' upside down. The Coat Hanger is one of those chance alignments of nature, as none of its ten stars form physical pairs or groups. The observer will revisit this asterism again and again, delighting in its beauty and humour. It is not however a naked eye object and is not even plotted in *Norton's Star Atlas*, although the excellent *Sky Atlas 2000* plots it in its entirety. The Coat Hanger is at its best when seen with binoculars, where it is revealed as a bright company of dazzling stellar points.

The deep sky showpiece of Vulpecula however, is the stunning planetary M27, otherwise known as the Dumb Bell Nebula. This is an incredible sight in binoculars, a blue white sphere of bright light trapped against a backdrop of hazy Milky Way stars in a fine field of luminous points. M27 can be found in the centre point of a group of five stars forming a W shape that can be perceived through a finder scope, lying above the tip of Sagitta, the constellation immediately to the south of this part of Vulpecula. Through a small telescope M27 shows itself to be a ring of blue light, which fades away towards its equator, then builds up again to complete the ring shape.

Messier 27

M27 was discovered by Messier in 1764. It is an eighth magnitude smudge of light that has a very high surface brightness; in fact it is probably the most conspicuous type of this object in the entire sky. The nebula lies at a distance of 1360 light years, giving the gaseous shell an overall diameter of 2.5 light years, which is very large for such a bright planetary nebulae. The central star is a very interesting object as it is the only known white dwarf associated with a planetary nebula to be a member of a binary system. Its companion is a K type dwarf of 17th magnitude, whilst the central star itself is only 14th magnitude, so it is not easy to spot this remarkable but faint pair.

The position of Vulpecula against the star clouds of the Milky Way guarantees the presence of those beautiful deep sky objects, open star clusters. There are five such objects that will interest the casual observer, each of them has a distinct beauty that makes the time spent browsing through the sumptuous clouds of the Milky Way worthwhile.

The first such cluster is the amazing NGC 6940, at RA 20h 34m 36s Dec 28°18m, a collection of over 100 stars in a large and faint scattered group lying in the Milky Way northeast of the W- shaped asterism where M27 can be located. NGC 6940 cannot be seen well in ordinary binoculars, but a giant pair will bring this object to light far better than a small telescope, as its wide field is ideal for this loose cluster. Most of the stars range from magnitude 11 to 13, whilst the cluster has an integrated magnitude of 10.5. Seen through a small telescope, the Milky Way stars suddenly give way to a condensation of starry points, and as one uses averted vision, countless luminous gems flash in and out of view in a kaleidoscope of light and patterns.

NGC 6823 is a gathering of thirty stars of magnitude 11 to 12 in a compressed field of 10th magnitude haze, which observers with large telescopes may relish, as the faint nebulae NGC 6820 lies in the same field of view. This is not a brilliant object for a small telescope, and is all but invisible in binoculars, but finding such objects against the confusion of the Milky Way is good practise for your celestial navigation, and will enable the observer to increase their experience of stars and associated fields and configurations.

A near neighbour in a low power ocular is NGC 6834, RA19h 52m 12s Dec 29°25m, a much richer aggregation of fifty or so stars of 11th magnitude shining collectively at magnitude 9.5. Observers with large telescopes are once again favoured with an additional object, this time the planetary nebulae NGC 6842, a 13th magnitude halo of light slightly smaller than M57, the Ring Nebulae in Lyra.

The last of these five clusters is a pretty group surrounding the star 20 Vulpeculae and known as NGC 6885, a bright group of thirty-five stars easily visible in a pair of binoculars. In a telescope this 7th magnitude cluster is well shown, with little streamers and star chains towards the centre of this engaging object.

The constellation contains a few more choice clusters for those with large telescopes, but for the modestly equipped observer, simply sweeping the Milky Way in this sector of stardust can reveal many more starry clumps and knots of interest, one of which is a bright but small star cloud designated NGC 6815, centred on RA 19h 40m 54s Dec 26°51m; a blur of light in binoculars but not well seen in a small telescope due to its low surface brightness. In the meantime, keep an eye on the star patterns hoping to spot that elusive nova.

CHAPTER SIX
The Circumpolar Constellations

*What they are saying is
that there is life there, too:
that the universe is the size it is
to enable us to catch up.
They have gone on from the human:
that shining is a refection
of their intelligence. Godhead
is the colonisation by mind
of untenanted space. It is its own
light, a statement beyond language
of conceptual truth. Every night
is a rinsing myself of the darkness
that is in my veins. I let the stars inject me
with fire, silent as it is far,
but certain in its cauterising
of my despair. I am a slow
traveller, but there is more than time
to arrive. Resting in the intervals
of my breathing, I pick up the signals
relayed to me from a periphery I comprehend*

—R. S. Thomas

Chapter Six –
The Circumpolar Constellations

In Britain, though we are not fortunate enough to view the southern wonders of the Magellanic clouds or the Carina Nebulae, we do have a company of stars that the southern astronomers cannot view to advantage. These groups of stars are visible all year round to anyone living north of 40 degrees latitude, and are known as the circumpolar stars.

The circumpolar constellations, as their name suggests, revolve continually around the pole of the Earth's axis, and can be seen swinging through the sky as the seasons change. The constellation of Ursa Major can be found low on the northern horizon in winter, whilst in spring it moves around to the east then overhead during the summer months. During autumn, this group is found to the west, before completing the circle and moving back to the north. The movement of Ursa Major is important, as it is the major circumpolar constellation, and the other groups can be found by following imaginary 'lines' through its constellation stars.

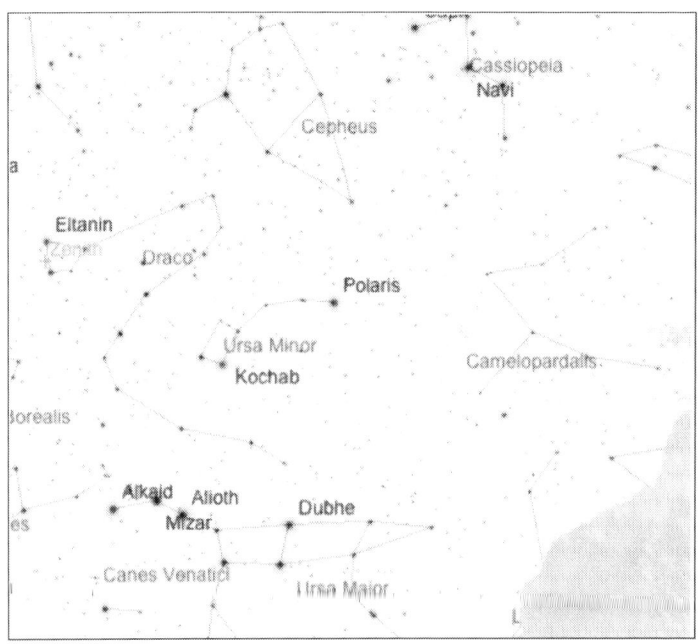

Circumpolar constellations

The other associations with which the observer will no doubt become familiar are Ursa Minor, Cassiopeia and Cepheus, whilst the other circumpolar constellations of Camelopardalis and Draco are not difficult to spot, but may take the beginner some time to trace out as they are rather winding or obscure.

The beauty of the circumpolar groups also lies in the fact that the Milky Way runs through two of the constellations of this part of the sky, namely Cassiopeia and Cepheus. Consequently, there is a lot on offer to the casual observer, not just once or twice a year but all the time! The star clusters of Cassiopeia are numerous, plus the double stars, coloured stars and Milky Way condensations make this constellation one of the most interesting groups of the sky. Additionally, the constellation of Cepheus contains the prototype of a giant variable star, an ideal stellar denizen from which to learn the vagaries of variable star observing. There is also the beautiful red giant μ Cephei, one of the largest stars known, plus a wealth of star clusters and even one or two galaxies.

Where would navigators be without Polaris? This is the brightest star near the northern celestial pole, lying less than one degree away and thus making polar alignment for northern observers very easy. Not for us the dim stars of the southern pole, rather, a wealth of constellations brimming with beautiful objects to entice and excite the observer.

If you are new to stargazing, get to recognize the circumpolar constellations until you know them like the back of your hand; they are the key to the rest of the sky visible from this country, and are precious groups in their own right. The circumpolar stars have a particular charm of their own and include some of the best and culturally important stars in the sky. The constellation of the bear has been accepted since ancient times, and its seven stars have been attributed mystical qualities by many cultures.

Distinctive Stars

Dubhe (spectral type K0II) and Merak (spectral type A1V) are known as the 'pointers' as they lie on the western edge of the 'bowl' of Ursa Major and a line drawn through them and extending northward brings the observer to the pole star, Polaris. Of the seven stars in Ursa Major, these two provide a nice contrast simply due to their different colour, as can be gauged from reading their spectral types above. Dubhe (Arabic for bear) has a seventh magnitude companion star, and although it is classed as an F8 white-yellow star observers have often remarked on its blue appearance when contrasted with the primary. Merak is the brightest member of a cluster called the Ursa Major Moving Group, or Collinder 285 and is about 80 light years away from Earth. Dubhe is not a member of this system, so comparing the two gives an impression of the three dimensionality of space.

The central star of the tail of the Bear is the famous double Mizar (spectral type A2V) and Alcor (spectral type A5V) that are well worth observing. One should separate them by using the naked eye only. A small telescope will reveal the system to contain four stars altogether, one bright companion close to Mizar, and a dimmer one between Mizar and Alcor.

Polaris (spectral type F8Ib) is a lovely star that marks the north polar axis of the Earth and is a boon to navigators. It is an extremely interesting star in that it was classed as a type of Cepheid variable star for many years, and then in the 1990s stopped pulsating altogether. Observers with binoculars may see a small circlet of stars of 7th and 8th magnitude just to the south of Polaris (depending on how the sky is oriented at the time). This asterism is known as the Engagement Ring, with Polaris set as the diamond. It is a rewarding sight. Polaris will remain the pole star for a short period of time on the astronomical scale; in 15,000 years time Vega in the constellation of Lyra will be the pole star.

One cannot mention the circumpolar stars without reference to one of the largest and most amazing stars in the sky, known only by its Greek letter μ. μ Cephei (spectral type M2Ia) is one of the largest red giants known, with an estimated diameter 1000 times that of the Sun. If brought to our solar system, even Jupiter would be orbiting within its envelope! It is an irregular variable too, with a mean period of 755 days. The great observer William Herschel called it the 'Garnet Star;' due to its red lustre and it is an easy object to spot visually as it lies just midway below a line drawn from δ Cephei to α Cephei. In binoculars it stands out as an orange-red dot whilst a small telescope may reveal more of its colour. The outer layers of μ Cephei are so cool that spectral bands of water vapour have been found. The distance to it is uncertain at present, but it is probably in excess of 1200 light years. Intervening dust clouds and gas probably attenuate the brightness of the star visually.

Many varied and wonderful stars that reside in other constellations accompany these beacons of the circumpolar night. Try discerning their colours and researching their types and other information. Doing so provides the observer with a holistic approach, and a deeper appreciation of the subject.

Camelopardalis
(Berecynthia)

Camelopardalis is an obscure train of stars that lies to the west of the constellation Ursa Major. It is not a constellation of antiquity, but its origins are somewhat clouded in mystery. It represents the giraffe, although commonly it has been associated with the 'ship of the desert', the camel, being placed on star charts by Bartschius in 1614 and representing the camel that brought Rebecca to Isaac in Hebrew tradition. The stars would have been included in the constellation of Berecynthia or Cybele by the ancient Welsh as they are rather obscure.

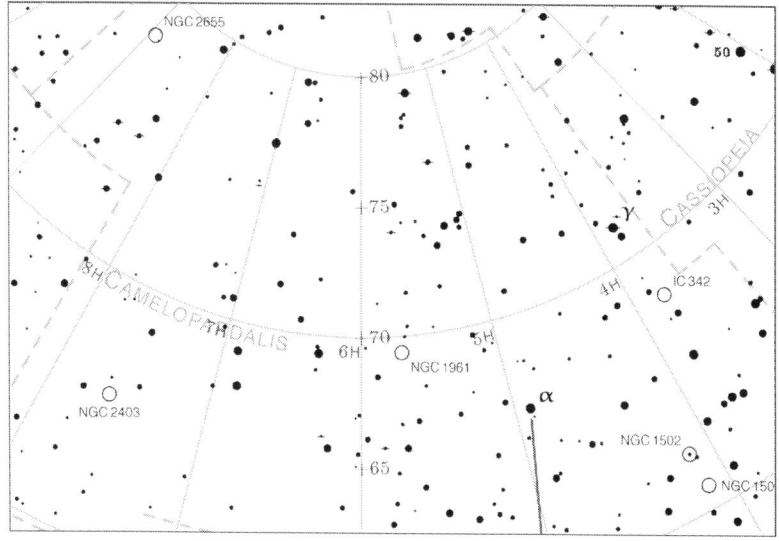

Camelopardalis

The two brightest stars of Camelopardalis can be found roughly 10 degrees above the eastern arm of Perseus. Both stars are of 4th magnitude, but the rest of the constellation spreads northwards and eastwards from these faint markers. Camelopardalis is a haven for faint galaxies, and the amateur with a large telescope will find rich pickings here. Otherwise, Camelopardalis contains little to interest the casual observer apart from the following objects.

One of the most fascinating stars in the sky is an 8.5 magnitude object called RU Camelopardalis. This apparently faint variable became one of the most observed stars in the sky for a period during the 1960s due to it being the only Cepheid type variable to ever stop pulsating. Formerly RU Cam varied between 8.3 and 9.2, but this period slowed, then stopped apparently for good, remaining at its current quoted magnitude. Once its position is known, it can be found with ease in binoculars or a small telescope. The star is an R type giant, one of a very rare class of 'carbon stars', as such it is a distinct orange-red in colour.

A lovely object for amateur scrutiny is the asterism Kemble's Cascade, a smattering of 5th magnitude stars best seen in winter. They lie above Perseus and arc across his head and shoulders in a line, leading to a small star cluster NGC 1502, a gathering of twenty-five 8th magnitude stars which contains two close binary systems that are a challenge for a small telescope. The cluster itself is fairly compressed; a high power will show it to better effect once the field has been found. Just to the north of this group at RA: 04h 07m Dec 60°55m is the beautiful ring like planetary nebulae NGC 1501. Not as bright or substantial as M57 in Lyra, it is nevertheless a great little object, shining at 11th magnitude as a wisp of blue white light.

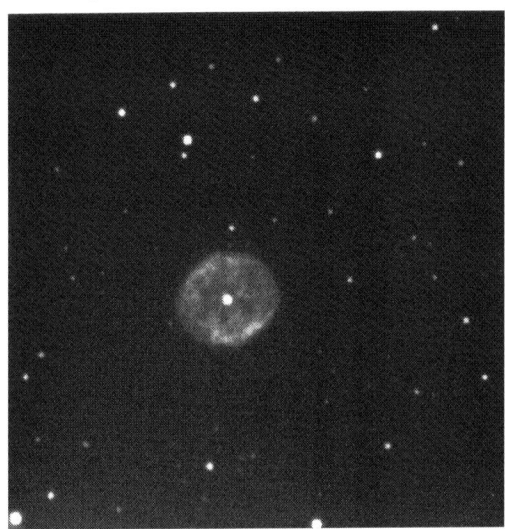

NGC 1501

The showpiece object of this obscure constellation is without doubt the Sc type galaxy NGC 2403 at RA 07h 36m 54s Dec 65°36m. This is one of a number of galaxies of the local group, almost the twin of the great M33 in Triangulum as it is remarkably similar in form. NGC 2403 is an 8th magnitude blob of blue-white light in a small telescope, whilst binoculars reveal it as a hazy patch of misty light that is easily mistaken for a comet. The galaxy appears to lie about 8 million light years away, and could be an outlying satellite of the M81-M82 group. NGC 2403 also is the first galaxy apart from M31 to have its distance measured by photographic studies of its Cepheid variables.

Cassiopeia
(The Goddess Don and Tale of Llyn Cwm Llwch)

This constellation represents Llys Don, the gateway to the fairy world and seat of Don and her son Gwydion, the King of the Fairies.

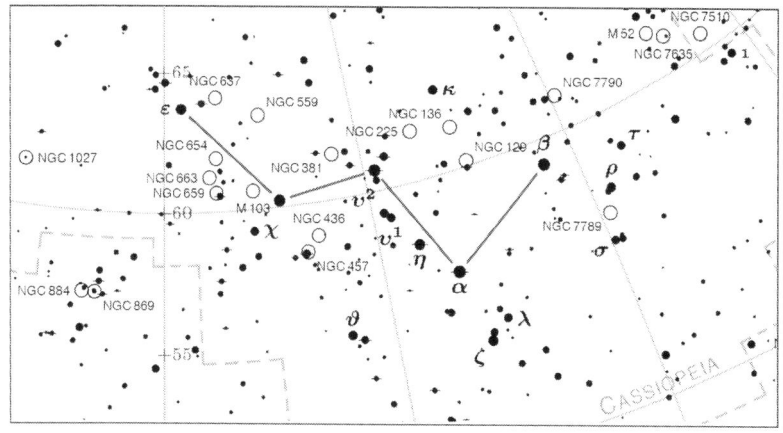

Cassiopeia

A Breconshire legend tells the tale associated with this constellation as it reaches its lowest point in the sky during the month of May. The story is set by the glacial lake at the bottom of Pen y Fan and involves an enchanted island which is generally invisible from the shore of Llyn Cwm Llwch. A passageway leading from the shore to this magical island would open up on May Day, the feast of Beltane, each year. Locals who had the courage to enter the doorway would find themselves in a beautiful garden on the island inhabited by fairies. The fairies would play wonderful music, tell stories, and give the visitors flowers and delicious fruits that appeared nowhere else in Breconshire. But these things came at a price – all the guests were warned that they must take nothing from the island.

But one May Day a guest put a flower in his pocket. When he emerged from the rock the flower vanished and he went mad. Furious with the theft, the fairy queen slammed shut the door, vowing never to open the passageway again. True to her word no one has since seen the magical island and fairy world of the May Queen. Nevertheless, anyone visiting Llyn Cwm Llwch can enjoy the tranquil beauty of the spot and wonder what hidden depths the lake contains.

* * *

Cassiopeia is a very conspicuous constellation, its five brightest stars making up a W shape in the circumpolar sky that can help the beginner to find other constellations around about. It lies in one of the richest portions of the Milky Way visible from Britain, and can be examined on any clear night all year round.

One of the finest star clusters in the Messier catalogue can be found in the Milky Way simply by extending a line through the two bright western stars of the 'W' out a similar distance as the gap between these two stars. There, in binoculars, should be a distinct glowing patch of stardust that a telescope will reveal as a bright compressed cluster of stars called M52. This beautiful cluster contains over 200 stars in a tight group of which the brightest members are blue and white giants shining with thousands of times the luminosity of the Sun. M52 is approximately 5000 light years away and is comparable in age to the Pleiades star cluster in Taurus.

Messier 52

A smaller cluster that is also a Messier object can be found close to δ Cassiopeia, but it is a rather unremarkable object in binoculars, and unfortunately not much better in a small telescope. This is M103, a cluster of forty stars lying in a very rich field of the Milky Way. M103 is very distant, over 8000 light years away, so if it were nearer it would be a far better object visually. Scanning the star clouds of the Milky Way around this object with binoculars or a small telescope can be very rewarding, as several clusters can be found very close by.

The best of these faint clusters is without doubt NGC 663, RA 01h 46m Dec 61°15m, a lovely congregation of about 60 stars of the 11th magnitude in a compressed group. NGC 663 lies at a comparable distance to M103, and in turn is enclosed by two fainter less rich star clusters namely NGC 654 and NGC 659, both of which contain around 25 to 40 stars each in a rich field.

NGC 663

The real prize of this constellation can be seen with a pair of good binoculars, but is absolutely spectacular in a small telescope. This is the star cluster NGC 457, the Owl cluster. In the eyepiece of a small instrument, NGC 457 is a blaze of 100 stars in the rough shape of this creature of the night. It can be located around the star ψ Cassiopeia, which forms one of the eyes of this beautiful object. Both the eyes appear to gleam in the darkness, and the body is outlined by fainter stars almost looking like ruffled feathers. Two chains of stars radiate out of the body in opposite directions, looking as if the owl is spreading his wings and about to fly, one of the few objects that really look like what their name portrays.

Another star cluster of note is the incredible NGC 7789, lying between σ and ϱ Cassiopeia, it is relatively easy to find its position, but can be difficult to spot due to its overall dimness, as it is found in an obscured area of the Milky Way. Yet once seen it will never be forgotten as NGC 7789 contains

NGC 457

one of the richest gatherings of stars in the sky. What a telescope reveals is a patch of stardust that scintillates like frost, over 900 faint stars flashing and shining tantalisingly, drawing the eye deeper into the recesses of space. NGC 7789 lies 8000 light years away in a dense part of the Milky Way. So rich is this cluster that, on long exposure photographs made by large observatories, astronomers have wondered if it was a depleted globular cluster. Studies have put it firmly in the classification of Galactic Clusters, but it is exceedingly rich even for these objects. Most of its stars are orange type giants, which are in the process of evolving away from the main sequence, indicating this cluster must be relatively old.

Contained in the southern portion of Cassiopeia are two faint galaxies, NGC 185 and NGC 147. These are of elliptical type and studies have shown that they belong to our local group; in fact they are outlying satellites of the great Andromeda galaxy M31. Both are rather faint objects, and may not be seen with a small telescope. However, a pair of giant binoculars may serve to study them as their low surface brightness may lend itself to the wide field capabilities of such instruments. Both galaxies lie over 2.2 million light years away and are composed of stars similar to those found in the globular clusters of our own galaxy.

One of the most massive stars in our galaxy can be found in Cassiopeia. This is the apparently faint 6th magnitude star AO Cassiopeia. It is a binary star with a total mass for the system of 80 times that of our Sun. One component has 23 solar masses and the companion has 57 solar masses. AO Cass is over 6000 light years away and probably shines with a luminosity of 300,000 times that of the Sun in total. The two components of the system are almost in contact with one another, so the stars are egg-shaped rather than round, and are also surrounded by tenuous shells of gas.

Within Cassiopeia are several good examples of nebulae, but unfortunately are not easy objects to view with average amateur equipment. The beginner would best be encouraged to scan a good star atlas or app and look for obscure clusters, which at least should be visible in small telescopes rather than these faint wisps of light.

Cepheus
(The God Beli)

The circumpolar constellation of Cepheus is not an easy group to define, as most of its stars are dimmer than one would expect of a constellation lying along the Milky Way. Cepheus lies close to the feet of his wife, Cassiopeia, a rather fitting aspect, as he appears to be rather a henpecked figure in mythology. He is the king of the Ethiopians and one of the participants in the drama of Perseus as father of the beautiful Andromeda.

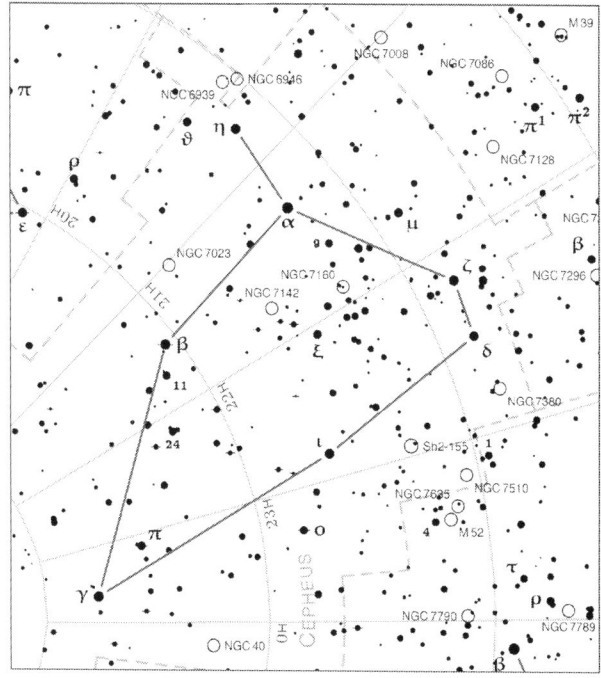

Cepheus

In the tales of Wales, Beli was founder of the Gwyr y Gogledd or the Men of the North, the husband of Don (Cassiopeia) next to whom he sits in the sky. Beli is the father of Arianrhod and is the grandfather of the great hero Llew Llaw Gyffes. In the *Mabinogion*, most of the activity of the sons of Don happen in Gwynedd, where her brother Math is king. Math and his nephew Gwydion are responsible for the creation of Blodeuwedd and the wedding of Llew (see Andromeda). Beli was also the god of the dead and the keeper of the gate to the underworld though he is also linked with the Sun, as some sources claim that his name lives on in the word Beltane – May Day or the first day of Summer.

* * *

The best way to visualise this constellation is to think of it as a child's drawing of a house, with a large peaked roof, and four stars – α β δ ι that mark out the corners of the walls, with δ Cephei and its attendants marking out a little triangle in the bottom east corner, with δ at the apex. Seen in this way, the Milky Way becomes the front garden running along the foot of the group. And what a garden it is! Cepheus is packed with beautiful things to explore, which can be seen all year round both with binoculars and small telescopes, making the group a delight to observe.

The star that attracts most observers to Cepheus is the yellow giant star δ Cephei, the prototype of the pulsating variables. The pulsations arise from deep within the layers of the star. It is at the helium burning stage but is subject to the $\varkappa$ mechanism, where the layers overlying the star's core become opaque to radiation coming from helium fusion. This radiation becomes trapped within the star causing it to swell, until a point is reached where the layers expand and cool, allowing the heat and radiation to flow out of δ Cephei. Once cooling has been achieved, the layers resume their normal proportions, becoming denser as gravity compresses them around the core and starts the process again.

John Goodricke first noticed its light variations in the seventeenth century, but its usefulness as a 'standard candle' was revealed by Henrietta Leavitt, who, in conjunction with Harlow Shapley, promulgated the period-luminosity law, a giant stepping stone on the road to cosmic discovery. This law states that brighter a particular Cepheid type variable is, the longer its period of variability. There are many Cepheids with regular periods ranging from hours to over 50 days, but the ones with periods of a week or more are the really bright stars, giants that can be seen over vast distances. Thus if a Cepheid is seen to have a particular period, then its distance can be inferred by using the relationship to describe how bright the star is. Once this brightness is known, then its distance can be found with some accuracy using well-established physical laws. In this manner, the distance to the galaxies was first defined, and Cepheids are still used as reliable distance indicators even today.

δ Cephei lies 887 light years away and has a luminosity of 2000 times that of the Sun. It varies from magnitude 3.6 to mag 4.3 in just over five days, so its periods are easy to follow. It is an ideal star to begin your variable star observations with, as it is visible all year round.

The other noteworthy star of Cepheus is μ Cephei, Herschel's 'Garnet Star'. It is found in the south of the constellation, at the baseline of the 'house'. It is an intense red in colour and is easily visible as a crimson glowing coal of the fourth magnitude in binoculars or a small telescope. μ Cephei is of spectral type M2Ia and is also an irregular variable, generally remaining around the fourth magnitude, but its light amplitude is not large. It is one of the largest red giant stars known, and one of the coolest.

When at minimum light, it has a temperature of only 1500 Kelvin, and its spectrum is a confusion of broad bands of molecules, including that of water vapour, one of only three stars in which such bands have been detected. μ Cephei is very remote, approximately 6000 light years away. It lies in an abundant field of stars on the edge of a large nebulae, I.C.1396, but the association is coincidental. This beautiful star is one of the treasures of the heavens, a lovely respite from the blues and whites of the mainstream of celestial lights.

One of the oldest, if not *the* oldest star cluster in our galaxy can be found in the northern part of this constellation. NGC 188, RA 00h 44m 18s Dec 85°21m, is a vast amalgamation of 300 faint stars, mostly of old red or yellow giants which are very difficult to detect in a small telescope. NGC 188 is found in the halo of our galaxy in contrast to other clusters, which prefer the company of the disc of the Milky Way. The group lies over 5000 light years away, and its brightest members shine feebly at 11th magnitude.

A contrasting star cluster can be found closer to the Milky Way. NGC 6939 is a compact gathering of over 100 stars which have an integrated magnitude of 9.5 at coordinates RA 20h 31m 24s Dec 60°38m. If you are viewing this group with a small telescope, then alter your sights a little to the southeast to find a small splodge of blue-white light called NGC 6946. This is an

Sc type galaxy just outside the local group at a distance of 12 million light years. The field of your eyepiece will be filled with stars as the Milky Way butts in a little to this rich region. The dissimilarity between these two objects makes a nice touch, leaving a distinctive memory of this unparalleled deep sky marvel.

There are several small star clusters within the little triangle of stars around delta Cephei that are worth investigating with binoculars or a small telescope. The best of these is the cluster NGC 7235, RA 22h 12m 36s Dec 57°17m, a group of 10th magnitude stars in a compact smattering of about twenty-five stars. Amongst this group can be seen the faint tendrils of the wispy Bubble Nebula. Another nice cluster worth seeking out is NGC 7142, RA 21h 45m 54s Dec 65°48m; an aggregation of fifty faint stars of around 11th magnitude, visible as a hazy spot of needle-points of light. NGC 7762 at coordinates RA 23h 49m 40s Dec 68°02m is worth watching out for, a compressed rich group of seventy faint stars that may be visible in a pair of giant binoculars. All these clusters are found along the richness of the Milky Way, so they may be difficult to spot amongst the background stars.

One gem of an object can be found at RA 21h 01m 35s Dec 68°10m 10s. This is the lovely reflection nebula NGC 7023, a dark patch in the Milky Way that resolves itself into a blue reflection around a central star. Indeed for this reason it is known as the 'Iris Nebula' after the flower. The area is easy to photograph and the nebula comes out very well.

NGC 7023

One of the bright stars of the group, γ Cephei at magnitude 3.2, makes interesting observing as it is one of the few naked eye stars which have a planetary system in evidence. The star is a K2V main sequence orange star with a mass similar to that of the Sun. The planet is 1.5 times Jupiter's mass lying 1.5 AU from γ, orbiting in 902 days. The whole system is only 30 light years away and would make an excellent target for exploration one day.

From a dark sky site, this particular stretch of the Milky Way glows with the star dust of countless celestial citizens, coloured stars and double stars can be found in great profusion here, and the observer can spend hours scanning this region of space night after night.

Draco
(The Mare of the Night and the Dragon of Wales)

In Welsh mythology Draco represents the horse that Berecynthia rode around the sky in her search for souls to steal. As she was not adept at telling the difference between sleep and death she would check each sleeper by having the horse breathe its foetid breath into the face of the person. If dead, they would be undisturbed but if asleep they would have vivid and scary dreams and upon waking they would realize that they had a visit from the night-mare, or as we would put it today, had a nightmare.

Draco also features in one of the greatest of Welsh myths, the one behind the dragon of the Welsh flag. The king of Britain, Vortigern, was building a castle on the hill known as Dinas Emrys near Beddgelert in north Wales, but the castle repeatedly collapsed due to earthquakes. Vortigern turned to his wise men who told him that he would have to find and sacrifice a fatherless boy to stop the earth shaking. Vortigern searched the country for such a rarity and found the young magician Merlin. About to sacrifice him, Vortigern was stopped when Merlin told him why the castle fell down. He said that under the foundations were two dragons who fought daily. Vortigen should clear the foundations, let the dragons fight and then use the victor as the signifier of his coat of arms.

This Vortigern did, and upon clearing the castle they discovered a white dragon and a red one. In a lengthy fight the red dragon killed the white one. Freed from its subterranean prison the red dragon flew into the heavens and Vortigern used the red dragon, Y Ddraig Goch, on his shield.

The dragon was also used by the British high king Cadwaldr, who was king of Gwynedd from 655-682AD. In Geoffrey of Monmouth's history of the kings of Britain, Cadwaladr is the last one to hold the title "high king of Britain" though he abdicated his kingship to go on a pilgrimage to Rome and the Holy land, as a prophet had told him that upon renouncing his power, the Britons would vanquish the Anglo-Saxons at some future time. This prophecy of deliverance has been held dear by many Welshmen since, chafing under the English yoke. It is little wonder that it became the nation's symbol.

Cadwaldr's heraldic Red Dragon bears more resemblance to the modern version of the Welsh dragon. During the Wars of the Roses, both Lancastrian and Yorkist houses used the Red Dragon of Cadwaladr as both claimed descent from the last high king of Britain. With the house of Tudor successful after the defeat of Richard III, perhaps many felt that this old prophecy was fulfilled now that the Welsh born Henry VII was king. Notwithstanding its origin, this representation from the night sky has now become the Dragon of Wales and is rampant on the flag of the country.

* * *

Draco is a long, sinuous constellation of rather faint stars that winds between Ursa Major and Ursa Minor and ends in a head of four stars close to Vega. The star α Draconis, or Thuban, is not the brightest star in the constellation but is remarkable in that 4300 years ago it was the closest star to the pole. A modern myth has it that within the great pyramids at Gizeh, several passages appear to be oriented towards this star's culmination, as seen from Egypt at that time. It has often been claimed that by looking up one of these passages, Thuban may be glimpsed in daylight by a keen eyed observer, although all stars, even of the first magnitude cannot be seen in daylight without some kind of visual aid. Thuban is otherwise undistinguished, being an A type star lying 303 light years away.

There are only a few objects worth the trouble to discover in Draco. One of these is a bright planetary nebulae, NGC 6543, the famous Cat's Eye Nebula at RA 17h 58m 36s Dec 66°38m which lies close to the pole of the ecliptic. This nebula is a difficult object to see with a pair of binoculars, as it is rather small in diameter, indeed, even in a telescope it looks like an out of focus star. With higher powers it can be seen as a bluish disc of 7th magnitude light, very pale in hue and unremarkable to the eye. In larger telescopes a distinct helix shape may be glimpsed under good conditions with the colour changing to a pale green.

The nebula is not particularly difficult to find. If you draw a line from δ to ζ Draconis, NGC 6543 lies midway along this line. It is mainly notable for being the first such object to be examined through a spectroscope. William Huggins (1824-1910) examined NGC 6543 and as a result was able to explain that far from being an unresolved cluster of stars, this nebula was in fact a cloud of luminous, rarefied gas.

Considering Draco's position far from the Milky Way, one would expect to find a profusion of galaxies to excite the observer. Unfortunately, there are few really bright galaxies here worth finding in small equipment, with the exception of NGC 5907, an Sb type which can be seen as an edge-on needle of grey light of 11th magnitude in a small telescope at coordinates RA 15h 15m 54s Dec 56°19m, and NGC 4125 an elliptical galaxy of similar magnitude to NGC 5907. Both objects are not good telescopic objects, and are virtually invisible to binoculars.

One of the nicest binocular doubles is the star ν Draconis in the head of the dragon. Both stars are of a similar magnitude at 4.8 but although both stars have spectral types of A6 and A4, they look distinctly green to most observers.

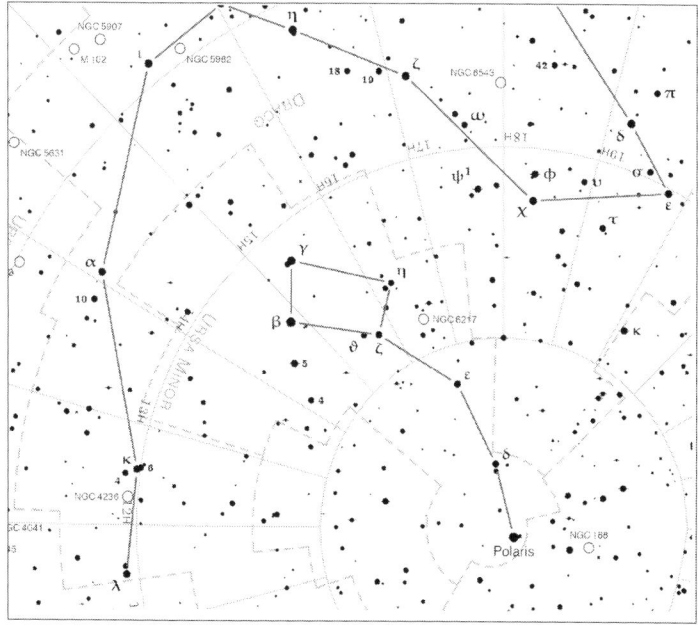

Draco

Draco has little else of note to the observer with limited equipment, though it does contain a few faint binary systems that may be captured by sweeping the constellation. The most interesting of these is possibly μ Draconis, a system of two red dwarfs in orbit around their common centre of gravity. Their orbital period is very long, over 4000 years, and at this time the two stars are drawing apart. There is increasing evidence of a third dwarf member, close in to the primary, but the system shows only two stars to observers with telescopes.

Ursa Major
(Berecynthia and the Plough)

This group of stars is probably one the best known constellations in the night sky. Since man first began to group the stars into some order this figure has been recognised as a bear, eternally drifting around the celestial pole. In Britain it is commonly called the 'Plough' after the tale of Hu Gadarn, who is extending his arm and holding the handle of the plough, the rest of the asterism crossing the sky through Auriga (the Yoke) across to Taurus the bull, pulling the plough. From Wales it can be seen on every clear evening as the major constellation of the circumpolar stars. Every kind of civilization has legends regarding it, from the Chinese to the Native Americans. Ursa Major is one of Ptolemy's original constellations, and is an ideal tool that will enable the beginner to navigate the northern sky.

The constellations of Ursa Major and Ursa Minor are easily recognizable

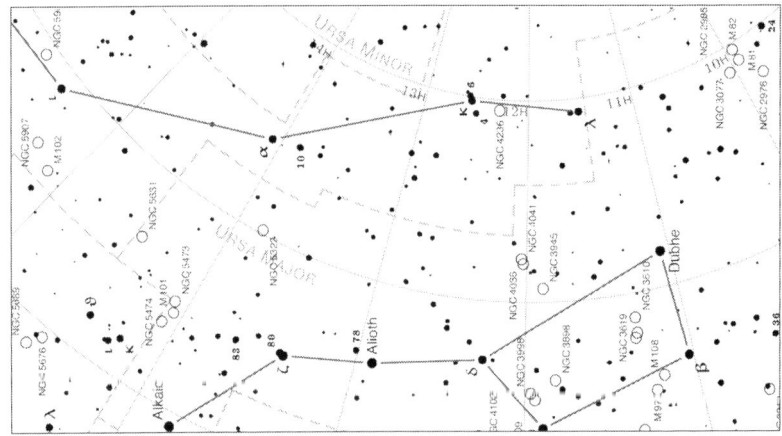

Ursa Major

to most people. The seven stars of Ursa Major make an asterism that point directly to the pole star Polaris; this sequence of stars being used by navigators for thousands of years. Beside being part of the legend of Hu Gadarn, the constellations of Ursa major and the twisting shape of Draco the dragon also play a part in Celtic legends.

The goddess Berecynthia, or Cybele, was identified by the Celts as the constellation of Ursa Major in pre-Roman times. This goddess of agriculture had a darker side however. When night fell, she would wheel about the sky riding her fearsome steed, Draco the Dragon. In Welsh tales the dragon became her bloody, brooding horse that Berecynthia rode each night in search of souls to steal. The old, young and the sick were killed in their beds (dying in their sleep in effect) to satisfy the craving for sacrifice that would yield new crops, whilst the living within the household would be troubled by vivid dreams and fitful sleep (see Draco).

* * *

Ursa Major consists of seven bright stars, five of which share a common motion through space due to their membership of the Ursa Major moving group, the closest star cluster known. Seventeen members of this moving cluster are bright naked eye objects, but there are up to fifty others occupying a region of space over 150 light years in extent. This greater area of stars, with roughly common motions is known as the Ursa Major Stream, and includes our Sun, although the proper motion of the solar system rules it out as being part of the cluster proper. Computations of the proper motions of the five cluster stars of Ursa Major reveal that the constellation will change very little over the next 250,000 years.

Of this group, the star Mizar is a beautiful introduction to double star observing, as even with the naked eye you can discern its little companion Alcor – The Rider. In a small telescope this system resolves itself into a collection of four stars gravitationally bound together, a very appealing sight to witness. The constellation is dripping with some of the most wonderful deep sky objects in the heavens, and as it lies quite far from the Milky Way, Ursa Major includes a variety of objects from planetary nebulae to galaxies.

A remarkable star is the faint Groombridge 1830, a dwarf type orange star shining at magnitude 6.5. It is one of the closest stars to Earth and is a fine example of a halo star, one different from the young objects of the galactic disc. Groombridge 1830 has one of the highest space velocities of any star: in 100,000 years it will be in the southern constellation of Lupus, invisible from

Britain. Most stars would take millions of years to make such progress. Groombridge 1830 is thought to have been one of the earliest types of stars to form within our galaxy as it condensed from a primeval gas cloud into stars. Groombridge 1830's frugal rate of nuclear burning has led it to lead a sedate life despite its high space velocity.

One of the showpiece objects in the northern sky is that of the galaxies M81 and M82. They are very bright, both shining at about 8th magnitude and easily visible in a pair of binoculars as two elongated patches of bluish light. In a small telescope they are transformed into two halos of white light, with M81, an Sb type, showing condensations and the hint of spiral arms, whilst M82 is a needle-like spindle of light with a distinct dark smudge at its centre. M81 is a Sa pec. type galaxy

Messier 81 and Messier 82

The surrounding field is full of faint stars, and over to the west of the two galaxies may be spotted a third by keen-eyed observers: this is a companion in their particular group, catalogued as NGC 3077, an E2 type galaxy. The three objects lie just outside the boundaries of our galaxy's local group, about 12 million light years away. M81 is one of the densest galaxies known, and may be harbouring vast amounts of dark matter in an invisible halo around it. A supernova exploded in this galaxy recently, becoming visible to amateur observers as a 10th magnitude speck of light, proving that constant observation will bring rewards.

Another beautiful galaxy worth finding is the elusive M101, an Sc type lying just above a trail of stars leading eastward from Mizar. It is more likely to be seen with a pair of good binoculars than a poor telescope, but once found it is a barely discernable glow of light almost merging with the background darkness. On a long exposure photograph, M101 is a stunning galaxy, a whirl of white arms and star fields that is a glory to behold. Its mottled blue-white shade in the eyepiece of a small scope gives just the tiniest hint of its full splendour. M101 lies outside our local group of galaxies at a distance of 20 million light years.

To the south east of the star β Ursa Majoris, is a faint but spectacular object called M97 the Owl Nebulae. This denizen of the sky is a 12th magnitude patch of glowing gas, a planetary nebula of amazing outline. Although it may seem to be dim from the magnitude given, it can be found with no difficulty at all with a 4" reflector. In such an instrument it looks like a fuzz of blue light, with a darker centre seemingly hiding the central star. This dark centre will reveal itself to larger telescopes as two distinct round patches, the eyes of the owl. Additionally, the bridge of the nose of the owl will show the dim 14th magnitude central star to a careful observer. However, the nebula is one of the most beautiful planetaries in the sky, an all-year wonder to those who live in Britain. M97 lies approximately 2000 light years away.

Messier 97

If you are using a low power ocular to view M97, then just over a degree away from the Owl ebulae is a lovely edge-on spiral, NGC 3556, which has come to be known as M108. It is a cigar-shaped needle of light shining at 9th magnitude with a enthralling electric blue glow due to the myriads of O and

B type giants in the edge-on arms of this captivating galaxy. NGC 3556 lies at a distance of 14 million light years from us.

One curious object in Ursa Major is the asterism M40. This is not a star cluster or a nebula, but a little group of two faint stars lying just above δ Ursa Majoris. It was noted by Messier after other observers drew his attention to it, and he recognised it as simply two stars, but included it in his catalogue under this number. It is not a remarkable object; indeed you would not glance at it twice.

Observations at Lick observatory in California have identified two Jupiter-like planets orbiting the star 47 Ursa Majoris. They have almost the same mass ratio as our own Jupiter and Saturn and travel in nearly circular orbits at distances far beyond that of the orbit of Mars around the Sun. These similarities suggest that this system formed in a similar way as our own solar system and theorists suspect that low mass, Earth-like planets might exist around 47 Ursa Majoris in its habitable zone. This star is also relatively bright enough to be seen with the naked eye. A small telescope will reveal its yellow colour and the mind can ponder the existence of a solar system remarkably like our own.

George Gatewood of the University of Pittsburgh has also discovered two Jupiter-like planets orbiting Lalande 21185. This interesting star is one of the closest to the solar system; a red dwarf just 8.25 light years away. It requires a telescope to view this dim red cinder, but its position is marked on *Sky Atlas 2000* for those who wish to pick it out with binoculars or a small telescope.

Ursa Minor (Berecynthia)

A famous constellation dating from antiquity and one of Ptolemy's original groups, although its identification as a bear can be explained mostly by its resemblance to Ursa Major. To the ancient Welsh it was part of the larger constellation of Berecynthia and her horse. The sky is full of animals that seem to come in pairs: dogs, bears, lions and even centaurs. Ursa Minor is an interesting constellation even though it contains little of note to the deep sky observer. Its salvation lies in it being the closest constellation to the north polar axis of the Earth's rotation, and a convenient point for navigational reference.

Once again, almost every civilization has identified this constellation as a bear, with the interesting exceptions of the great Asian cultures in India and Cambodia. This is probably due to their rise as city states when the constellation

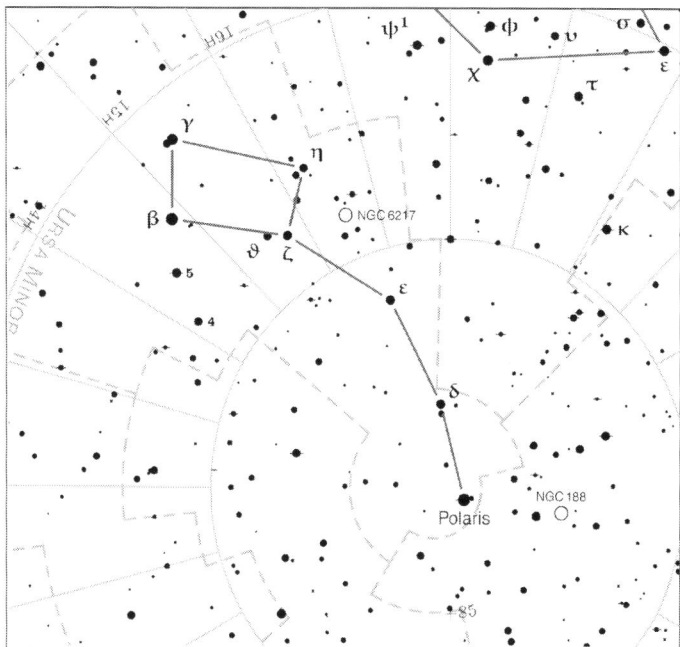

Ursa Minor

assumed its present position 2500 years ago. To these enlightened societies, the constellation was the Celestial Mountain or the Mountain of the World, and several Hindu and Khmer temples, such as Kandarya Mahedevi and Angkor Wat were erected as towering earthly representations of this group of stars.

The pole star, Polaris, has not always held this position. As previously noted, in ancient Egypt, α Draconis, was the closest star to the pole. This shift has come about due to the precession of the Earth, like the wobbling of a child's spinning top as it rotates. In 25,000 years time, the pole star will be the beautiful Vega, and the current summer constellations will become circumpolar.

Polaris is an F7 spectral type with a pale yellow hue and is the 49th brightest star in the sky. It has a faint companion of ninth magnitude lying quite close to the primary. Polaris is a relatively distant star, 325 light years away and shining with a luminosity of 2500 times that of our Sun, but is slightly variable, the variations being of small magnitude over a period of 3.96 days.

The second brightest star in Ursa Minor is Kocab, which is the brightest star in the square of the body of the bear. Its name is an Arabic derivation for 'Guardian of the Pole'. Only 3000 years ago it was the closest bright star to the true pole, closer than Polaris will ever come. Its beauty lies in its lovely orange colour, which is immediately evident to the observer with binoculars. Kocab is a K type giant star, over 130 light years away and has an output of 400 times that of our Sun.

Ursa Minor is found in a rather sparse region of stars far from the glories of the Milky Way. It has however several faint galaxies that observers with large telescopes may like to try to discover.

I stood and stared; the sky was lit,
The sky was stars all over it,
I stood, I knew not why,
Without a wish, without a will,
I stood upon that silent hill
And stared into the sky until
My eyes were blind with stars and still
I stared into the sky.

—Ralph Hodgson

Conclusion

It is hoped that the reader is well on their way to identifying and enjoying the constellations and their associated deep sky objects. No matter whether you are equipped with binoculars, a small telescope or simply your naked eye, the pleasure of observing and having an intimate and holistic knowledge of the starry heavens can only add to your appreciation of the fantastic universe in which we live, of which our Earth is a mere speck lost in the isolation and vastness of space.

For those who wish to pursue their interest at an amateur level, it is recommend that you join a local astronomy club, or even the British Astronomical Association, whereby you can have encouragement and instruction from experienced observers and teachers, in fact a broad spectrum of astronomical life. Let them advise and guide you through the pitfalls that the beginner usually undergoes, learn from their mistakes, and those of your own, and make your hobby something to be proud of, and even more important, something that you enjoy.

Several international societies provide some great resources for those who wish to expand their observing skills or concentrate on a selected class of object through personal interest. One of the best of these is the *Astronomical League*, which although primarily American, one can join for a small fee and become a 'member at large'. The League offers a variety of observing challenges that can culminate in obtaining a certificate of recognition for one's work. This gives one confidence in one's skills and ability to use equipment, as well as giving impetus to your observing projects. Such awards as the Messier, Herschel, Lunar, Planetary Nebulae and Star Cluster certificates, along with many others the League offers, prompt a good reception amongst observers in astronomical societies and of course promote one's expertise and growth in your subject. The league can be explored at: www.astroleague.org/

In the UK the British Astronomical Association has several observing sections that promote differing agendas and all observations are gratefully received and correlated. They have deep sky, planetary, photography, historical and instrument sections and more. They also run schools and conferences that enable the novice to learn more and to improve their skills at many different levels. Details of these can be found online at http://www.britastro.org/

Another fine observing society that concentrates on deep sky objects and produces an excellent journal is the Webb Deep-Sky Society; a group that concentrates on producing excellent handbooks, star charts and much more for general use. They too have an annual conference which makes anyone interested in astronomy very welcome. They can be found online at www.webb deepsky.com/

One of the largest amateur conferences in Europe is the annual *Astrofest* held in Kensington, in London, on the first weekend in February. This is not just a series of talks by astronomy experts but is a marketplace for the latest in astronomical equipment and imaging and has many stalls selling items, generally at discounted prices. Additionally, there is the International Astronomy Show held at the Warwickshire Exhibition Centre at Leamington Spa.

There are several fields of study and research that are truly valuable in which amateurs make a vital contribution. Nova and supernova patrol is one area in which amateurs are taking the lead. Many discoveries made by amateur astronomers alert the professionals to transient phenomena. There is an ongoing need for observers to be alert for the long overdue supernovae that could appear at any time in our home galaxy, and many amateurs are flocking to these growing ranks, performing vital work that the professionals do not have the time or equipment for. Many are pushing the limits of current technology in an effort to record and photograph elusive celestial visitors such as Near Earth Objects and integrate their data into an international watch on such celestial hazards. Others are monitoring the skies for comets and other transient phenomena. In observational astronomy, one can never be bored!

Whatever you decide to pursue, remember this: enjoy it! The universe is not something that can be fully understood by superlatives, by numbers or formulae, or by concentrating on specifics; it is something to be sensed, experienced, savoured and valued.

Everyone has a personal history that motivates one to learn about the night sky, so do not be put off by the cold, the dark, one's inexperience or even the intemperance of others. The night sky has an immediacy that supplants every thought and fills the mind with great emotion; in fact, there is nothing quite like a perfect night under the stars to make you as close to nature as possible.

Glossary

Absorption
Absorption is a property of atomic elements in that they absorb a photon of light of a particular wavelength, resulting in the electron(s) within the atom either jumping to a higher orbit in the atom (excitation) or leaving the atom altogether, a process known as ionization. This leads to the development of a dark line in the spectrum of a star or other body at the specific energy or wavelength of the absorbed photon.

Absolute Magnitude
The apparent magnitude that a star would possess it if were placed at a distance of 10 parsecs from Earth. In this way, absolute magnitude provides a direct comparison of the brightness of stars. The apparent magnitude of a star is based upon its luminosity and its distance. If all stars were placed at the same distance then their apparent magnitudes would only be dependent on their luminosities. Thus, absolute magnitudes are true indicators of the amount of light each star emits.

Accretion
An accumulation of dust and gas into larger bodies such as stars, planets and moons, or as discs around existing bodies.

Albedo
A measure of the reflectivity of an object and is expressed as the ratio of the amount of light reflected by an object to that of the amount of light incident upon it. A value of 1 represents a perfectly reflecting (white) surface, whilst a value of zero represents a perfectly absorbing (black) surface. Some typical albedos are: The Earth – 0.39; The Moon – 0.07; Venus – 0.59.

Aphelion
The point in an orbit around the Sun at which an object is at its greatest distance from the Sun (opposite of perihelion).

Apogee
Similar to aphelion. The point in an orbit when a body orbiting the Earth, (eg Moon or artificial satellite.) is farthest from the Earth (opposite of perigee).

Apoapsis
The point in an orbit when a planet is farthest from any body other than the Sun or the Earth.

Arc Minute
A measure of angular separation – one sixtieth of a degree or one sixtieth of an hour of right ascension.

Arc Second
Another measure of angular separation – one sixtieth of an arc minute (1/3600th of a degree).

Glossary

Ascending Node
The point in the orbit of an object when it crosses the ecliptic (or celestial equator) whilst moving south to north.

Asteroid
(Also 'planetoid') These are rocky bodies, the vast majority of which orbit the Sun between Mars and Jupiter. It is thought that there are around 100,000 in all. The largest asteroid is Ceres, which has a diameter of 1000 km. The smallest detected asteroids have diameters of several hundred feet. Together with comets and meteoroids, asteroids make up the minor bodies of the solar system. They are considered to be the left-over planetesimals from the formation of our solar system. The gravitational pull of Jupiter is thought to have stopped the members of the asteroid belt from forming a planet.

Astronomical Unit. (A.U.)
The mean distance from the Earth to the Sun, i.e. 149,597,870 km.

Aurora
A glow in the Earth's ionosphere caused by the interaction between the Earth's magnetic field and charged particles from the Sun (the solar wind). It gives rise to the 'Northern Lights', or Aurora Borealis, in the Northern Hemisphere, and the Aurora Australis in the Southern Hemisphere.

Baader Astro Solar Film
A neutral density film that reduces the intensity of sunlight by 99.99%, allowing direct viewing through an appropriate telescope.

Bessell Filters
The generally used UBVRI photometric system of colour filtration applied to CCD photography.

Binary Star
A system of two stars orbiting around a common centre of mass due to their mutual gravity. Binary stars are twins in the sense that they formed together out of the same interstellar cloud.

Blue Moon
The second full moon in a calendar month, or the third full moon in a season containing four.

Broadband Filter
A filter that is generally used to reduce light pollution as it transmits the wavelengths of light for $H\alpha$, OIII and $H\beta$ but stops the transmission of light wavelengths inimical to sodium and mercury vapour streetlights.

Caldwell catalogue
A catalogue of 110 objects constructed by the British amateur astronomer Patrick Moore and based on the famous Messier catalogue by the eighteenth century French observer Charles Messier. The Caldwell catalogue is named after Patrick Moore's surname; which

was the hyphenated Caldwell-Moore. It contains objects from the NGC and IC catalogues and covers both southern and northern celestial hemispheres.

Celestial Equator
The projection of the Earth's equator upon the celestial sphere.

Celestial Sphere
The projection of space onto the night sky, an imaginary hollow sphere of infinite radius surrounding the Earth but centred on the observer. First postulated by Ptolemy. It is the basis of sky charts, and the celestial co-ordinate system. The coordinate system most commonly used is right ascension and declination. The sphere itself is split up into arbitrary areas known as constellations.

Celestial Poles
The projection of the Earth's poles onto the celestial sphere.

Chromosphere
The layer between the photosphere and the corona in the atmosphere of the Sun, or any other star, mainly composed of excited hydrogen atoms. In a Hα telescope the chromosphere appears to have a myriad of bright points across the solar disc, a phenomenon known as the Chromospheric Network.

Coma
(1) The dust and gas surrounding the nucleus of a comet.
(2) A defect in an optical system which gives rise to a blurred, pear shaped, comet-like image.

Comet
An icy object in independent orbit about the Sun, smaller than a planet and usually presenting a highly elliptical orbit extending out to beyond Jupiter.

Conjunction
When two bodies appear to close together in the sky, i.e. they have the same Right Ascension. Mercury & Venus are said to be at superior conjunction when they are behind the Sun, and at Inferior conjunction when they are in front of it. The outer planets are simply said to be at conjunction when they pass behind the Sun.

Constellation
An arbitrary grouping of stars that form a pattern. The sky is divided into 88 constellations. These vary in size and shape from Hydra, the sea monster, which is the largest at 1,303 square degrees, to Crux, the cross, which is the smallest at 68 square degrees.

Corona
The outer layer, and hottest part, of the Sun's atmosphere.

Coronagraph
A special telescope which blocks light from the Sun's disc, thus creating an artificial eclipse, in order to study its atmosphere.

Cosmic Ray
An extremely fast, energetic and relativistic (high speed) charged particle.

Cosmos
The Universe: the word is derived from the Greek, meaning 'everything'.

Crater(s)
A depression in the lunar or planetary surface caused by an impact from a large meteor or asteroid. Generally circular in appearance and occasionally marked with a central peak and collapsed walls.

Culmination
An object is said to culminate when it reaches its highest point in the sky. For northern observers, this occurs when the object is due South. For southern observers when it is due North.

Declination
A system for measuring the altitude of a celestial object, expressed as degrees north, or south, of the celestial equator. Angles are positive if a point is north of the celestial equator, and negative if south. It is used, in conjunction with Right Ascension, to locate celestial objects.

Descending Node
The point in the orbit of an object, when it crosses the ecliptic whilst travelling north to south.

Direct Motion (Prograde Motion)
(1) Rotation or orbital motion in an anticlockwise direction when viewed from the north pole of the Sun (i.e. in the same sense as the Earth); the opposite of retrograde.
(2) The east-west motion of the planets, relative to the background of stars, as seen from the Earth.

Digital Camera
This can be the Single Lens Reflex camera (SLR) which instead of having a standard film inside now relies on an imaging chip to capture the subject in the same manner as a video camera or Charge Coupled Device (CCD) camera. It can also refer to any compact digital camera that uses chip technology and to differentiate them the larger SLR types are known as DSLR for shorthand.

DMK Camera
A camera that uses digital technology to capture image files in the form of a movie that can then be downloaded and stacked in an appropriate software programme such as Registax. They are used for lunar, solar and planetary imaging.

Dwarf Star
A star that lies on the main sequence and is too small to be classified as a giant star or a supergiant star. For example, the Sun is a yellow dwarf star.

Eccentricity
The eccentricity of an ellipse (orbit) is the ratio of the distance between its foci and the major axis. The greater the eccentricity, the more 'flattened' is the ellipse.

Eclipse
A chance alignment between the Sun, or any other celestial object, and two other celestial objects in which one body blocks the light of the Sun, or other body, from the other. In effect, the outer object moves through the shadow of the inner object.

Ecliptic
The apparent path the Sun (and, approximately that of the planets) as seen against the stars. Since the plane of the Earth's equator is inclined at 23.5 degrees to that of its orbit, the ecliptic is inclined to the celestial equator by the same angle. The ecliptic intersects the celestial equator at the two equinoxes.

Ellerman Bombs
Microflares in the solar chromosphere associated with magnetic field reconnections, where two opposing streams of ionized material collide with a brief flare of light and energy. A small solar flare.

Elongation
The angular distance between the Sun and any other solar system body, usually the Earth, expressed in degrees. The term Greatest Elongation is applied to the inner planets, Mercury and Venus. It is the maximum elongation from the Sun. At greatest elongation, the planet will appear 50% phase.

Emerging Flux Region
Areas on the sun where a magnetic dipole, or flux tube, is surfacing on the disc and can produce a bipolar sunspot group.

Ephemeral regions
Limited energy magnetic dipoles with lifetimes of about a day that contain no sunspots. Ephemeral regions can develop anywhere on the Sun, but are more common at mid and lower solar latitudes.

Equatorial Mount
A telescope mount designed so that the two axes, which support it, are aligned one to the polar axis and the other to the Earth's equator. Once an object is centred in the telescope's field of view, only the polar axis need be adjusted to keep the object in view. If the polar axis is driven at Sidereal rate, it will counteract the rotation of the Earth, keeping the object (except the Moon) stationary in the field of view.

Equinox
This is the time when the Sun crosses the celestial equator. There are two equinoxes; Vernal (Spring), around March 21st and Autumnal (Autumn) around September 23rd. On these dates, day and night are equal. Actual dates and times vary due to the Earth's precession.

Glossary

Faculae
Unusually bright spots, or patches, on the Sun's surface. They precede the appearance of sunspots and can remain for some months afterwards.

Fibrils
Fine structure in sunspot areas associated with spicules and solar activity in the chromosphere.

Filament
A strand of (relatively) cool gas suspended over the Sun (or star) by magnetic fields, which appears dark against the disc of the Sun. A filament on the limb of the Sun seen in emission against the dark sky is called a solar prominence.

Galaxy
Vast star systems containing thousands of billions of stars, dust and gas, held together by gravity. Galaxies are the basic building blocks of the Universe. There are three main classes, Elliptical, Spiral and Barred, named after their appearance.

Galilean Moons
Jupiter's four largest moons: Io, Europa, Ganymede and Callisto. First discovered by Galileo.

Geosynchronous Orbit
Sometimes known as a geostationary orbit, in which a satellite's orbital velocity is matched to the rotational velocity of the planet. As such, a geostationary satellite would appear to be stationary relative to the Earth.

Globular Cluster
A spherical cluster of older stars, often found in galaxies.

Granulation
The mottled, orange peel, appearance of the Sun's surface; caused by convection within the Sun.

Gun Griz photometric system
A photometric calibration system for professional use which is referenced with known stars of particular spectral character and brightness.

Heliocentric
Sun-centred system of cosmology.

Hypersensitize
The process of treating a photographic film with hydrogen or nitrogen forming gas to render the emulsion more sensitive to light and to reduce reciprocity failure with long exposures

Inclination
(1) The angle between the orbital plane of the orbit of a planet, and the ecliptic.
(2) The angle between the orbital plane of a satellite and the equatorial plane of the body it orbits.

Inferior Conjunction
When Mercury, or Venus, are directly between the Sun and Earth.

Inferior Planets
These are the planets Mercury and Venus. They are called inferior planets because their orbits lie between that of the Earth and the Sun.

Interstellar Medium
The material that fills the voids in space between the stars. The ISM is mostly hot vaporous gas such as coronal gas and stellar winds but does contain hydrogen and helium left over from the big bang, in addition to elements seeded into the ISM by the death of stars in planetary nebulae or supernovae.

Light Year
The distance travelled by light in one year, equal to 9,460,712,000,000 km.

Limb
The outer edge of the disc of a celestial body.

Luminosity
Absolute brightness. The total energy radiated into space, per second, by a celestial object such as a star.

Luminence Layer
The image taken by a CCD camera through a hydrogen alpha, SII or CaII filter which is then added to an BVR image to gain maximum input from the astrophysical image.

Lunation
The period between successive new moons.

Magnetosphere
The region of space where a planet's magnetic field dominates that of the solar wind.

Magnitude
The degree of brightness of a celestial body designated on a numerical scale, on which the brightest star has magnitude -1.4 and the faintest star visible to the unaided eye, has magnitude 6. A decrease of one unit represents an increase in apparent brightness by a factor of 2.512. Apparent magnitude of a star is the brightness as we see it from Earth, whilst absolute magnitude is a measure of its intrinsic luminosity. Lower numbers represent brighter objects.

Mare
Areas on the lunar surface that were once thought to be seas of water. (Hence mare – Latin for sea.) Any open surface on a planet that is a lava plain.

Meteor
Also known as a 'shooting star' or 'falling star', is a bright streak of light in the sky caused by a meteorite as it burns up in the Earth's atmosphere.

Meteorite
A rock of extraterrestrial origin, found on Earth.

Minor Planets
Another term for asteroids.

Moon
A naturally occurring satellite, or relatively large body, orbiting a planet.

Mylar filter
A solar filter that allows less than 1% transmission of light through a metalized filter to enable safe solar viewing in white light.

Nebula
A term used to describe celestial objects which have a fuzzy, or nebulous, appearance (from the Latin for cloud), but now used to describe clouds of gas or dust that have condensed out of the Interstellar Medium (ISM).

Nebula Filters
Generally wide bandpass filter or light pollution filter that allows the passage of Hα, OIII and Hβ wavelengths through to a camera, optical system or CCD camera.

Nova
An existing star that suddenly increases its brightness by more than 10 magnitudes and then slowly fades. Novae are generally associated with binary stars in which one of the stars is a white dwarf in close proximity to the primary star. The primary star sheds gas to the white dwarf that allows build up on the surface until pressure and temperature ensure a huge thermonuclear detonation.

Occultation
This is when one celestial body, passes in front of, and obscures, another.

Open Cluster
A group of young stars, possibly bound together by gravity, that formed together.

Opposition
A planet is said to be 'in opposition' when it appears opposite the Sun in the sky. For the outer planets, this is generally the closest they come to the Earth, hence when they are most easily visible.

Optical Binary
A pair of stars that happen to lie close to one another on the celestial sphere because of a chance alignment. They are not physically associated with one another and lie at vastly different distances. Optical binaries are also known as visual binaries.

Orbit
The path of one body around another due to the influence of gravity.

Parallax
The angular difference in apparent direction of an object seen from two different viewpoints.

Parsec
A unit for expressing large distances. It is the distance at which a star would have a parallax of one arc second, equal to 3.2616 light years, 206,265 astronomical units (AU) or 30.857 * 10E12 km.

Penumbra
Means, literally, 'dim light'. It most often refers to the outer shadow cast during eclipses, and defines the region of shadow which gives rise to a partial eclipse. It is also the lighter area surrounding the central region of a Sunspot.

Periapsis
The point in an orbit closest to a body other than the Sun or the Earth.

Perigee
The point in its orbit where the Moon, or planet is closest to the Earth.

Perihelion
The point in its orbit when an object is closest to the Sun.

Perturb
To cause a celestial body, to deviate from its predicted orbit, usually under the gravitational influence of another celestial object.

Photosphere
The visible surface of the Sun.

Plage
Bright regions in the Sun's chromosphere.

Planisphere
An aid to locating stars and constellations in the night sky. It consists of two discs. One with the entire night sky, and the other, which covers the first, having a window through which a portion of the sky can be seen. The second disc is set according to the date and time.

Precession
Circular motion about the axis of rotation of a body; fixed with respect to the stars. The Earth is a giant gyroscope whose axis passes through the North and South Poles and this axis precesses with a period of 27,700 years.

Prominence
A cloud, or plume, of hot, luminous gas in the solar chromosphere. It appears bright when seen against the cool blackness of space. When they are in silhouette against the disc they are known as filaments. They are mainly composed of hydrogen, helium and calcium.

Glossary

Quadrature
When a superior planet; Jupiter, Saturn etc.; is at right angles to the Sun, as seen from Earth.

Quasars
Compact, extra galactic, objects at extreme distances, which are highly luminous. They are thought to be active galactic nuclei. The name is an acronym for 'quasi-stellar radio source'. A quasar is very similar to a QSO (quasi-stellar object) but gives out radio waves also.

Radiant
The part of the sky from which a particular meteor stream appears to come from. Meteor showers are usually named after the constellation in which the radiant originates.

Red Giant
A spectral type K or M star nearing the end of its life having a low surface temperature and large diameter, e.g. Betelgeuse in Orion..

Red Shift
The lengthening of the wavelength of electromagnetic radiation caused by relative motion between source and observer. Spectral lines are red-shifted from distant galaxies, indicating that the galaxies are moving away from us due to the expansion of the Universe.

Resolution
The amount of small detail visible in an image (usually telescopic). Low resolution shows only large features, high resolution shows many small details.

Retrograde
Rotation of a planet, or orbit, opposite to that normally seen.

Right Ascension (RA)
The angular distance, measured eastwards, from the Vernal Equinox. It is one of the ordinates used to reference objects on the celestial sphere. It is the equivalent to a longitude reference on the Earth. There are 24 hours of right ascension within 360 degrees, so one hour is equivalent to 15 degrees. Together with declination, it represents the most commonly used co-ordinate system in modern astronomy.

Semi-major Axis
The semi-major axis of an ellipse (e.g. a planetary orbit) is half the length of the major axis which is a segment of a line passing through the foci of the ellipse with end points on the ellipse itself. The semi-major axis of a planetary orbit is also the average distance from the planet to its primary.

Sidereal Time
Star time; the hour angle of the vernal equinox. Time measured with respect to the fixed stars rather than the Sun.

Sidereal Month
The 27.32166 day period of the Moon's orbit.

Solar Continuum Filter
A green light filter transmitting light wavelengths centred at 510nm rendering a visible green image of the sun. Such filters are used in conjunction with either a Herschel Wedge or Baader astro filters

Solar Cycle
The 11-year variation in sunspot activity.

Solar Flare
A sudden, short-lived, burst of energy on the Sun's surface, lasting from minutes to hours.

Solar Wind
A stream of charged particles emitted from the Sun that travels into space along lines of magnetic flux.

Solstice
This is the time when the Sun reaches its most northerly or southerly point (around June 21st and December 22nd respectively.). It marks the beginning of Summer and Winter in the Northern Hemisphere, and the opposite in the Southern Hemisphere.

Spectral Classification
A method of classifying stars based upon the appearance of hydrogen absorption lines in their spectra. The spectral sequence OBAFGKM was determined by Williamina Fleming and Annie Jump Canon in the early twentieth century.

Star Cluster
A loose association of stars within the Milky Way. Examples are the Pleiades (Seven Sisters) and Hyades clusters.

Sunspot
A cooler region of the Sun's photosphere (which, thus, appears dark) seen as a spot, on the Sun's disc. They are caused by concentrations of magnetic flux, typically occurring in groups or clusters. The number of sunspots varies according to the Sun's 11-year cycle. More sunspots are seen at the Maxima of solar cycles, with few being observed during the Minima between.

Superior Conjunction
This is when Mercury, or Venus, is behind the Sun.

Superior Planets
Also known as the outer planets. These are the planets beyond the Earth's orbit. They are, in order: Mars; Jupiter; Saturn; Uranus; Neptune; Pluto.

Supernova
An exploding star, usually quite massive in comparison to the Sun. There are two main types of supernova; Type Ia are white dwarf stars that exceed the Chandrasekhar mass of 1.4 times that of the sun, whilst Type 1B Ic and Type II are massive stars that explode once the iron fusion stage is reached.

Glossary

Terminator
The boundary between day and night regions of the moon's, or a planet's, disc.

ToUcam
A small webcam that fits in the eyepiece holder of a telescope to gain a direct video image of an astronomical object. Manufactured by Phillips.

Transit
The apparent journey of Mercury or Venus across the Sun's disc, or of a planet's moon across the disc of its parent.

UBVRI
The coloured filter photometric system generally employed by amateur astronomers and systematized by Michael Bessell in the 1990s taken from original work by Johnson and Cousins in the 1950s and 1960s.

Umbra
From the Latin for shade, it is the shadow area defining a total eclipse. Or the dark central region of a sunspot.

Unsharp masking
A photographic and image reduction technique that allows the stacking of many images to gain increased detail and resolution in an astronomical object.

Variable Star
Any star whose brightness or magnitude varies with time. The variations can be intrinsic because of internal processes or extrinsic, due to eclipses, dust and other phenomena. Variations can also be irregular or periodic.

White Dwarf
A whitish star, of up to 1.4 solar masses, and about the size of the Earth with consequential very high density, characterized by a high surface temperature and low brightness.

Wratten Filters
Coloured glass filters with a range across the visible spectrum from red to blue that enable the blocking of partcular longpass wavelengths of light in order to see more detail on planetary and lunar surfaces. They are indicated by particular numbers that are standardized across a range of colours.

Zenith
The point on the celestial sphere directly above an observer, or the highest point in the sky reached by a celestial body.

Zenithal Hourly Rate (ZHR)
It is the number of meteors per hour, for a particular stream, that is estimated will be seen under favourable seeing conditions if the radiant were directly overhead the observer. Usually the actual figure is less than this.

Zirin Class
The different active or quiescent features of prominences in the solar chromosphere developed by Harold Zirin.

Zodiac
The apparent path, in the sky, followed by the Sun, moon and most planets, lying within 10 degrees of the celestial equator. Ancient astrologers (nothing to do with modern astronomy!) divided it into 12 groups, the Signs of the Zodiac, though there are actually 13 astronomical constellations which lie on the zodiac, since the Sun passes through Ophiuchus each December. Ophiuchus is not recognized by astrologers.

Zodiacal Light
A faint glow from light scattered off interplanetary dust in the plane of the ecliptic.

Bibliography

Bayley M. *Caer Sidhe* Vol 1 and Vol 2 Capel Bann Publishing 1997

Becker A. & Noone K. *Welsh Mythology and Folklore in Popular Culture* McFarland Press 2011

Davies S. *The Mabinogion* Oxford University Press 2007

Glyn-Jones K. *Messier's Nebulae and Star Clusters* Cambridge University Press 1974

Kerrigan M. *Celtic Legends. Heroes and Warriors, Myths and Monsters* Amber Press 2016

Miles-Watson J. *Welsh Mythology. A Neostructuralist Analysis* Cambria Press 2009

O'Meara S. *The Messier Objects* Cambridge University Press 2013

Rhys J. *Celtic Folklore, Welsh and Manx* 1901

Rolleston T. *Myths and Legends of the Celtic Race* Read Books 2013

Skene W. *The Four Ancient Books of Wales* Oxford University Press 1868

Watkins G. *Welsh Legends and Myths* Lulu Publishing 2012

Index

Achernar 100
Actaeon 99
Adar Llwch Gwin 169-70
Addanc 132, 182, 202
Adhara 82
Agammenon 173
Albireo 75, 167, 174, 214
Alcock, GED 181
Alcor 220, 235
Alcyone 124
Aldebaran 21, 71, 82, 123, 125, 130
Algol , 69186
Alnilam 117
Alnitak 117, 118, 119
Alpha Persei Moving Group 85
Altair 21, 162, 164, 170
Alt-azimuth mount 25-6
Anaxagorus 83
Andromeda 42, 45-9, 63, 64, 69, 74, 76; α Andromedae 45; β Andromedae 45; γ Andromedae 45; υ Andromedae 48-9; Rho Andromedae 48; Sigma Andromedae 48; Theta Andromedae 48; R Andromedae 48; 13 Andromedae 48
Andromeda (myth) 58, 62, 227
Aneirin 7
Annwn 32
Antares 22, 44, 71, 132, 163, 164, 203-4
Aperture Index 24
Apollo 31
η Aquarids 33-4
Aquarius 36, 42, 43, 49-53; Beta Aquarii 50; R Aquarii 53
Aquila 42, 162, 163, 164, 167, 169-72, 178, 200, 213; α Aquilae (Altair) 170; γ Aquilae 171
Aratus 92, 110
Arawn 66
Arcturus 21, 128, 130, 131, 132, 133
Argo Navis 119
Argo Navis 120
Arianrhod 30, 31, 65, 91, 140-1, 228
Aries 42-43, 53-5; γ Arietis 53-5; 30 Arietis 53; 53 Arietis 55
Arion 179
Artemis 86
Arthur 7, 103, 119, 145, 185, 199-200, 207
Asterion 133
Atlas 124
Auriga 78, 80, 85, 86-90, 110, 123, 130, 234; AE Auriga 55, 89-90; 18 Auriga 85; 19 Auriga 85, 234
Aurora 34-35
Aurora (goddess) 34
Aurora Australis 35
Aurora Borealis 35
Averted Vision 22, 74

Baade Walter 169
Barnard's Loop 119
Barnard's Star 189
Bayer, Johann 17, 104
Beddgelert 94
Bede 12
Bedivere 200
Bedwyr 72, 95
Beehive Nebula (M44) 92
Belenos 123
Belenus 31, 62
Beli 31, 62, 228
Bellatrix 117
Beltane 31, 104, 223, 228
Berecynthia 133, 221, 235
Berecynthia (myth) 231, 233
Berenice 136
Bessel, Friedrich 96, 177
Betelguese 22, 79, 81, 86, 117, 132, 164
Binoculars 23-25
Al Biruni 83
Black Book of Carmarthen 7-8, 56
Black Eye Galaxy (M64) 138
Balack Hole: Cygnus X1 178
Blinking Nebulae (NGC 6826) 1786
Blodeuwedd 30, 45, 66, 75
Blue Flash Nebula (NGC 6905) 181
Bode, Johann 138

Index

Bolide 33
Boötes 21, 123, 128, 131-33; ε Boötes 132-3; T Boötes 133
Borrow, George: *WildWales* 9
Bran 31, 142
Branwen 31, 142
Branwen, Daughter of Llyr 15
Brecon Beacons Observatory 9, 32, 162
Bridal Veil Nebula (NGC 6992) 177
Bridgid 156-57, 173-4
Brochi's Cluster 167
Brochwel 107
Brychan 100
Bubble Nebula (NGC 7235) 230
Burnham's Celestial Handbook 43

Caelum 80
Cai 72, 95
California Nebulae 69
Camelopardalis 219, 221-23; RU Camelopardis 221-22
Cancer 90-94, 113, 128; α Cancri 93-4; Zeta Cancri 94
Canis Major 78, 79, 80, 81, 82, 86, 94-8, 119; γ Canis Majoris 98; δ Canis Majoris 97; τ Canis Majoris 86, 97; Sirius B 96-97
Canis Minor 21, 78, 79, 81, 98-100; β Canis Minoris 100 (Gomeisa)
Canes Venatici 36, 103, 131, 133-6; α Canum 135; β Canum 135
Cannon, Amy Jump 20
Canopus 130
Cantre'r Gwaelod 182
Capella 80-1, 87
Capricorn 43, 55-7, 70, 71; α Capricorni 57; β Capricorni 57-8; γ Capricorni 58; 41 Capricorni 58
Caput 163, 211
Carina 121
Carina Nebulae 218
Carn Gefallt 95
Cassiopeia 18, 47, 67, 166, 188, 219, 223-7; δ Cassiopeia 224; σ Cassiopeia 225; ψ Cassiopeia 225; AO Cassiopeia 225
Cassiopeia (myth) 58

Castor 82, 104, 111, 173
Caswallawn 140
Cat's Eye Nebula (NGC 6543) 233
Cauda 161, 211
Ceffyl-Dwr 61-3
Celtine 146
Cemaes Head 179-80
Centaurus 129, 159
Cepheus 94, 108, 166, 219, 227-31; α Cephei 218; γ Cephei 231; δ Cephei 166, 220, 228-9; μ Cephei 219, 220, 229
Ceres 156
Ceridwen 30-1, 32
Cernunnos 115
Cerridwen 141
Cetus 42, 58-62; τ Ceti 61; O Ceti 59, 60 (Mira)
Cetus (myth) 58
Chara 133
De Cheseaux, PL 205
Chiron 194
Christmas Tree cluster (NGC 2264) 86
Clark, Alvan 80, 96
Clouds: B59 168; B65 168; B66 168; B67 168; B72 (S Nebulae) 192; B78 (Pipe Nebulae) 192; B111 167; B112 167; B119 167; B133 171; B143 171
Clytemnestra 173
Coathanger group 167
Collinder 70 cluster 86
Columba 80; Mu Columbae 55
Coma Berenices 128, 129, 136-40
Corona Borealis 65, 128, 129, 140-1, 148, 178, 201; α Coronae (Gemma) 141; ρ Coronae 141; T Coronae 141; R Corona Borealis 141
Corvus 128, 142-4, 152; δ Corvi 143
Crab Nebulae (M1) 36, 52, 125-6
Crannog 49
Crater 128, 144-46, 152; γ Crateris 146; R Crateris 146; U Crateris 146
Creidyladd 11, 148
Crux Scutum 167
Culhwch 95, 115, 193
Culhwch and Olwen 15, 54, 72, 95, 103,

115, 119, 192
Custennin 54
Cwm Corlwyd 72
Cwn Annwn 56, 103, 133, 148
Cwn Wybir 133
Cybele 221
Cybele (myth) 235
Cygnus 75, 81, 162, 166, 172-8, 184, 188, 199, 213; α Cygni (Deneb) 174; β Cygni (Albireo) 174, 214; γ Cygni (Sadir) 175; η Cygni 178; χ Cygni 178; P. Cygni 178, 201; 16 Cygni 177; 52 Cygni 177; 61 Cygni 177; HD 226868 178
Cygnus Rift 83, 167, 171, 178, 210
Cyledr 103

Dafydd Meurig 116
Daghda 193
Dagon 55, 70
Dark nebulae 39
St David 186, 211
Delta 94
Deneb 81, 162, 164-5, 166, 167, 174-5
Delphinus 111, 162, 179-81; γ Delphini 180; R. Delphini 181; HR. Delphini 181
Democritus 83
Dinas Rock 116
Dobsonian mount 26
Don 30, 140, 223, 228
Draco 150, 211, 219, 232-4, 235; α Draconis (Thuban) 232, 239; δ Draconis 233; ς Draconis 233; μ Draconis 234; V Draconis 233
Draco (myth) 234
Dreyer, JLE 36-7, 52
Drudwas ap Tryffin 170
Drudwyn 95, 115, 193
Druids 14-15, 91, 123, 173-4
Dubhe 219-20
Dumb Bell Nebula (M27) 215
Dylan 30, 91

Earth 31-2, 33, 34
Efnisien 142
Efrawg 112, 203

Einion 100-1
Emission nebulae 37
Enif 63
Eog 92
Epona 62
Epsilon 95
Equatorial mount 26
Eratosthenes 55
Erichtonius 86
Eridanus 78, 100-3; β Eridanis 100, 101-2; ε Eridani 102-3; O Eridani 102
Eridanus (myth) 100
Eskimo Nebulae (NGC 2392) 106
Eta 95
Europa 123
Eyepieces 28-9

Fish Mouth Nebulae 118
Flmaing Heart Nebulae (NGC 2024) 119
Flaming Star Nebulae 85
Flamsteed, John 18
Focal ratio 26
Fomalhaut 44, 70-1
Fomalhaut b 44
Fornax 42

Galaxies 35, 37
Galileo 83
Gatewood, George 238
Gawain 103, 170
Al Geidi 57
Gelert 94-5
Gemini 78, 82, 90, 99, 103-7, 110, 111, 148, 173; α Geminorum (Castor) 104; χ Geminorum 106; λ Geminorum 106; U Geminorum 106; 85 Geminorum 106
Geminids 33-4
General Catalogue of Nebulae and Star Clusters 37
Geoffrey of Monmouth 12, 228
Gerald of Wales 199
Globular clusters 38
Gomeisa 100
Goodricke, John 69, 228
Goronw 65-6

Goronwy 45, 75,
Gould, Benjamin 84
Gould Belt 84, 86
Great Nebulae (M31) 45
Y Groes Fendigaid 117
Groombridge 1830 233
Guest, Lady Charlotte 8, 95
Guinevere 31
Gunnion Fort 206
Gwalchmai 46, 169
Gwene 31
Gwern 142
Gwerne 30
Gwibr 150-1
Gwion 30-1
Gwrhyr 192-93
Gwyddno Garanhir 56, 182
Gwydion 45, 66, 82, 91, 140, 161, 169, 223, 228
Gwladys 100-1
Gwyllgi 98-9
Gwyn 103-4
Gwyn ap Nudd 32, 56, 148
Gwythyr 104

Halley, Edmund 18, 33, 146
Halley's Comet 33
Hebe 142
Helen of Troy 171
Helith 145
Helix Nebula (NGC 7293) 50-1
Hephaestus 84
Hera 81, 88, 144, 163, 171
Hercules 126, 127, 144-8; α Herculis 147-8
Hercules (myth) 81, 88, 144-5, 163, 192, 197
Herne the Hunter 113
Herschel, John 25, 37
Herschel, William 37, 38, 81, 106, 109, 111, 175, 195, 203, 218, 227
Hevelius 55, 108, 131, 205
Hind, JR 42, 106
Holy Grail 142-3
Hooke, Robert 54-5
Horse Head Nebula 117

Hu Gadarn 85, 121, 129-30, 131, 201, 209, 234
Hubble, Edwin 38, 46, 83, 113, 129
Hubble Classes 37
Huggins, William 233
Hyades 82, 93, 125, 126
Hydra 128, 146, 150-3
H20 cluster 201

I.C. 417 Nebula 89
I.C. 443 Nebula 89
I.C. 1396 Nebula 229
I.C. 4665 cluster 167, 191
I.C. 4756 cluster 167, 213

Idris Gawr 149, 150, 184-5
International Dark Sky Reserves 13

Jason 120
Jupiter 32, 94, 238
Jupiter (god) 32, 193

Kant, Immanuel 83
Keystone group 148
Kocob 240

Lagoon Nebula (M8) 168, 194-5
Lascaux caves 13-14
Leavitt, Henrietta 20, 228
Leda 173
Leo 71, 90, 115, 128-29, 152, 153-6, 158; α Leonis 154; γ Leonis 154; λ Leonis 156; R Leonis 154-5; 18 Leonis 154; 19 Leonis 154
Leonids 33, 34, 156
Lepus 78, 80, 107-8; α Leporis 109; β Leporis 109; R Leporis 108-9
Libra 181-84; β Librae (Zuben El Genubi) 182-3; δ Librae 183; HD 195019 181
Little Dumb Bell Nebula (M76) 69-70
Llamhigyn y Dwr 202
Llandegla 98
Llangorse Lake 49-52, 202
Llathen Teiliwr 117
Llew Llaw Gyffes 31, 45, 65-66, 75, 82,

109, 165, 169, 228
Llwybr Llaethog 167
Llyn Bala 32, 185
Llyn Bras 182
Llyn Cwm Ffynnon 202
Llyn Cwm Llwch 223-4
Llyn y Fan Fach 157
Llyn y Ffynnon Las 132
Llyn Lladaw 200
Llyn Llion 132
Llyn Llyw 72
Llyn Nad y Forwyn 136
Llyr 60, 142
Llywelyn the Great 94
Lud 32
Lughnasadh 31
Lupus 129, 235
Lynx 109-11, 153, 181; Alpha Lyncis 111; ψ Lyncis 111
Lyra 42, 148, 162, 184-8, 220, 222; β Lyrae 186; ε Lyrae 186
Lyrids 34

Mabinogion 8, 62, 72, 95, 112, 141, 169, 170, 192, 203
Mabon 72, 94-5, 115, 192-93
Maddolwch 142
Madoc 118-20
Markarian's Chain galaxy 159
Maksutov (telescope) 26
Manger Nebula (M44) 92
Maraldi, Giovanni 50
Marius, Simon 45
Mars 32, 238
Math 45, 66, 228
Mayor, Michele 65
Meara 99
Mechain, Pierre 36, 138, 155, 156
Medusa 58
Melangell 107
Melotte III 136
Melotte 15 cluster 85
Merak 219-20
Mercury 31
Merlin 145, 200, 231

Mermaid 179-80
Messier, Charles 36-7, 45, 47, 50, 51, 69, 109, 125, 159, 187, 215, 237
Messier Catalogue 29, 36, 37, 73
Messier Objects *M1* 35, 52; *M2* 35, 50-1; *M3* 35, 134; *M4* 204; *M5* 212; *M6* 205-6, 212; *M8* 168, 194-5, 199; *M9* 191; *M10* 189-91; *M11* 167, 208; *M12* 189-91; *M13* 198; *M14* 191; *M15* 64; *M16* 168, 212-13; *M17* 168, 197; *M18* 168; *M19* 192; *M20* 168, 195; *M21* 167; *M22* 195-6, 198, 206; *M23* 196-7; *M24* 168, 196, 197, 199; *M25* 197; *M26* 209; *M27* 69, 186, 215, 216; *M28* 196; *M29* 167, 175; *M30* 57, 58; *M31* 45, 48, 64, 223, 226; *M32* 46-7; *M33* 75, 76; *M34* 69; *M35* 85, 105; *M36* 85, 89; *M37* 85, 89-91; *M38* 85, 88, 89; *M39* 166, 175; *M40* 238; *M41* 80, 97; *M42* 54, 76, 86, 118, 194, 212; *M43* 86, 118; *M44* 91-2, 94; *M45* 69, 125; *M46* 86, 121-22; *M47* 86, 121-2; *M48* 152; *M49* 160; *M50* 113; *M51* 52, 135; *M52* 166, 224; *M53* 137; *M54* 197-98 *M55* 198; *M56* 185; *M57* 184-85, 220; *M59* 158; *M61* 160; *M62* 206; *M63* 135; *M64* 134; *M65* 155, 156; *M66* 155, 156; *M67* 93-4; *M68* 153; *M71* 200-1; *M72* 51; *M73* 51; *M74* 73-4; *M75* 198; *M76* 69-70; *M77* 59, 60; *M78* 118; *M79* 109; *M80* 204; *M81* 236; *M82* 236; *M83* 152-3; *M84* 159; *M86* 159; *M87* 159; *M88* 139-40; *M90* 160; *M92* 149; *M93* 135; *M95* 156; *M96* 156; *M97* 236-7; *M99* 138-9; *M100* 138; *M101* 238; *M103* 224-5; *M104* 135; *M106* 135; *M108* 237
Meteors 33-4
Milky Way 13, 37, 45, 46, 66, 67, 70, 74, 82-4, 86, 88, 89, 96, 98, 99, 100, 102, 104, 106-7, 111, 113, 115, 117, 121, 126, 129, 130, 132, 146, 149, 152, 155, 162-3, 165-9, 170, 173, 175, 177-8, 181, 184, 187, 188, 191, 192, 194, 196, 199, 200, 201, 204, 205, 206, 207, 209, 213, 215, 216, 219, 220, 225, 226, 227, 228, 230, 231, 233, 235, 240
Minerva 173
Mintaka 117

Mira 60-1
Mirak 132-3
Mirphak 67
Mizar 132, 220, 235, 236
Moon 30, 32
Monoceros 76, 86, 99, 112-15, 152; β Moncerotis 114; R Moncerotis 114-5
Morvugh 56

Nad the Maiden 136
Nebulae 36-8
Nennius 95, 205
Neptune 32
New General Catalogue 36-7
NGC Objects
147 226; *157* 60; *185* 226; *188* 94, 229; *205* 47; *246* 61; *404* 49; *457* 225; *488* 74; *604* 76; *654* 225; *659* 225; *663* 225; *744* 85; *752* 47; *869* 67-8; *884* 67-8; *891* 49; *908* 60; *957* 85; *1245* 69; *1501* 222; *1502* 222; *1528* 69; *1535* 102; *1545* 69; *1647* 126; *1746* 126; *1807* 126; *1817* 126; *1907* 89; *1931* 89; *1977* 118; *2017* 109; *2024* 119; *2158* 85, 105; *2237* 114; *2244* 86, 114; *2264* 86, 114; *2266* 106; *2301* 115; *2324* 115; *2354* 98; *2360* 98; *2362* 86, 97; *2392* 106; *2403* 222-3; *2419* 111, 180-1; *2420* 106; *2438* 121; *2506* 115; *2539* 122; *2683* 111; *2903* 156; *3077* 236; *3242* 152; *3556* 238; *3628* 154; *3887* 146; *4038* 144; *4125* 233; *4312* 135; *4361* 143; *4435* 159; *4438* 159; *4565* 139, 178; *4567* 1560; *4568* 160; *4594* 142, 144; *5053* 137; *5466* 132; *5694* 153; *5897* 183; *5907* 233; *6210* 149, 150; *6229* 149; *6267* 150; *6453* 206, 232-3; *6520* 198-9; *6522* 169; *6530* 195; *6535* 213; *6539* 213; *6543* 232-3; *6568* 198; *6572* 192; *6603* 196; *6611* 212; *6633* 191; *6645* 198; *6649* 210; *6664* 210; *6682* 210; *6683* 210; *6705* 188; *6709* 171; *6712* 209-10; *6755* 171; *6760* 171; *6781* 171; *6790* 171; *6791* 187-8; *6811* 176; *6815* 216; *6819* 177; *6820* 216; *6823* 216; *6826* 177; *6834* 216; *6842* 216; *6866* 176-7; *6871* 167; *6885* 216; *6905* 181; *6934* 180; *6939* 229; *6940* 216; *6946* 229; *6992* 177; *6997* 175; *7000* 166, 174; *7006* 111, 180-1; *7009* 51-52; *7023* 230; *7062* 175; *7082* 175; *7142* 230; *7235* 230; *7293* 52-3; *7331* 64; *7662* 48; *7762* 230; *7772* 64; *7789* 188, 225-6
Non of the Stream 188, 211
North America Nebulae 174
Norton's Star Atlas 37, 156, 214
Nova: Cygni 1975 178
Nudd Llaw Ereint 32
Nwython 103

Object Glass 24, 26
Offrwm 146, 147
Olwen 95, 115
Omega Nebulae (M17) 197
Ophiuchus 128, 148, 163, 178, 188-92, 201, 211; β Ophiuchi 192, 167; η Ophiuchi 191; θ Ophiuchi 192; ϱ Ophiuchi 168, 204-5; Lambda Ophiuchi 191; 30 Ophiuchi 191
Orion 15, 18, 21, 55, 78, 79, 81, 84, 86, 90, 94, 106, 113, 115-9, 194, 212; α Orianis (Betelguese) 117; β Orianis (Rigel) 100, 117
Orion's Belt 103, 118
Orion nebula 76, 117-8
Orionids 33, 34
Orpheus 184
Osiris 107
Owain ap Gruffydd 150-1
Owain Llaw Goch 116
Owl cluster (NGC 457) 225
Owl Nebulae (M97) 237

Pan 55
Parthenius 146
Payne-Gaposhkin, Cecilia 20
Peekaboo Nebulae 119
Pegasus 42, 45, 62-5, 72; ε Pegasi (Enif) 63; 53 Pegasi 65; ε Pegasi (Enif) 63; 51 Pegasi 65; HD 209458 65
Peredur 112, 203
Perseids 33, 34, 156

Perseus 42, 65-70, 78, 84, 164, 186, 221, 222; α Persei 69; β Persei (Algol) 69; χ Persei 84-5; λ Persei 84-5
Perseus (myth) 227
Phaeton 100
Phaeton 3200 33
Pickering, Edward 20
Pigott, Edward 138
Pinwheel Galaxy (M33) 75
Pipe Nebulae (B78) 168, 192
Pisces 42, 72-6; α Piscium 73; η Piscium 74; 107 Piscium 74
Pisces Austrinus 42, 44, 70-2; Pi Pisces Austrinus 72; Lacaille 9352 72
Planetary nebulae 38
Pleiades (M45) 13, 69, 97, 123, 124, 1253, 224
Plough 233-4
Pluto 32
Polaris 219, 220, 234, 235, 240
Pollux 82, 104, 173
Porrima 160
Poseidon 60
Praesepe Nebula (M44) 92
Procyon 21, 79, 81, 98, 99-100
Prydwen 119
Ptolemy 86, 98, 107, 120, 136, 234, 238
Puppis 78, 80, 86, 119-22; 19 Puppis 122
Pwca 58
Pwll 45, 62
Pwll, Lord of Dyfed 15
Pyrddyn 100-1
Prydwen 119
Pyxis 80

Quadrantids 34
Queloz, Didier 65

Radio galaxies 59
Rasalgethi 149-50
Realm of the Nebulae 129, 137-8
Red Book of Hergest 7-8
Reflection nebulae 37
Regulus 21, 71, 82, 130, 154
Rhiannon 43, 62

Rhiryd 119-20
Richard the Tailor 107-8
Rigel 21, 81, 82, 100, 102, 117, 130
Ring Nebulae (M57) 186, 216
Ringtail Galaxy (NGC 4038) 144
Rosette Nebulae (NGC 2244) 114
Rosse, Lord (William Parsons) 52, 125, 135

S Nebulae (B72) 192
Sadir 175
Sagitta 162, 163, 199-202, 215
Sagittarius 55, 57, 162, 163, 168, 169, 178, 192-9, 203, 204; γ Sagittari 178, 199; θ Sagittari 201; μ Sagittari 198; ς Sagittari 197; FG Sagittari 201; WZ Sagittari 201-2
Saiph 117
Saith Seren 124
Samhain 31
Sarn Gwydion 66, 82, 165
Saturn 32, 238
Saturn Nebulae (NGC 7009) 52
Schmidt Cassegrain 26
Scorpius 71, 132, 162, 163, 164, 168, 182, 192, 202-6; α Scorpii (Antares) 213
Sculptor 42
Scutum 162, 167, 168, 206-10; α Scuti 210; β Scuti 210; ε Scuti 209 (M26); R. Scuti 208-9
Scutum Sobieski 163, 207
Selkie 56
Serpens 128, 163, 211-113
Sextans 128
Seyfert galaxies 59-60
Sgwd Einion 100-1
Sgwd Gwladys 100-1
Shapley, Harlow 38, 84, 111, 149, 228
Siamese Twins galaxies NGC 4567 & 4568) 160
Silverberg, Robert 53
Sipher, Vesto 83
Sirius 21, 79, 80, 81, 95-7, 163-4, 170
Skull Nebula (NGC 246) 61
Sky Atlas 2000 37
Smyth, William 208
Sombrero Galaxy (NGC 4594) 143

Index

Spica 21, 130
Spiral nebulae 38
Star clusters 36
Star List 2000 43
Summer Triangle 162, 164
Sun 21, 31, 32, 86, 90, 129, 164
Swan Nebulae (M17) 197
Swift Tuttle 33
Sword Handle 67-8, 70

Taliesin 7, 30-2, 72, 115, 119; *The Spoils of Annwn* 119, 145
Tama Rereti 83
Taranis 32, 193
Taurus 13, 21, 36, 71, 82, 86, 93, 123-6, 131, 224, 234; Theta Tauri 125
Tegid 32
Telescopes 9, 25-8
Telyn Idris 184-5
Tempel-Tuttle comet 33, 156
Thubun 232
Tiamat 150
θ Trapezium 118
Y Tri Brenin 117
Triangulum 42, 43, 74-6; Iota Trianguli 75
Trifid Nebulae (M17) 195
Al Tusi 83
Y Twr Tewdws 124
Tylwyth Teg 103, 136
Twrch Trwyth 95, 115, 119, 153-4, 158, 185

Uther Pendragon 115
Uranus 32
Ursa Major 15, 86, 122, 129, 133, 218, 219, 221, 232, 234-8; β Ursa Majoris 237; δ Ursa Majoris 238; 47 Ursa Majoris 238; Groombridge 1800 235; Lalande 21185 238
Ursa Major Stream 96, 235
Ursa Minor 219, 232, 234, 238-40

Vela 121
Velpecula 69
Venus 30, 65
Vega 21, 88, 162, 163-4, 186, 220, 232

Vindemiatrix 160
Virgo 128, 128, 130, 132, 135, 138, 152, 156-60; ϱ Virginis 159; Gamma Virginis 160
Virgo Cluster 129, 139, 155, 184
Vortigern 231
Vulpecula 42, 69, 186, 200, 213-16; 20 Vulpeculi 216

Water Jar asterism 49
Webb, TW 73
Whirlpool Nebulae (M51) 135
White Book of Rhydderch 7-8
Wild Duck cluster (M11) 167
Witch Head Nebulae 102, 117
Wodliparri 83

Y Gododdin 7
Ysbaddaden 54, 95, 115
Ysbryn 146-7
Ysgydion 146-7

Zenithal Hourly Rate 33-4
Zeus 30, 123 144, 146, 165, 173, 179, 193, 194
Zuben El Genubi 182-90

About the Author

Martin Griffiths is an enthusiastic science communicator, writer and professional astronomer. Over his career he has utilized history, astronomy and science fiction as tools to encourage greater public understanding of science. He is a recipient of the Astrobiology Society of Britain's Public Outreach Award (2008) and the Astronomical League's Outreach Master Award (2010). He also holds the League's Master Observer certificate and has written or contributed to over 100 published science articles for many journals.

He was one of the founder members of NASA's Astrobiology Institute Science Communication Group, which was active between 2003-2006. He also managed a multi-million pound European programme in Astrobiology for adult learners across Wales in 2003-2008. Since then he has been involved in promoting adult education across Wales and after spending 17 years as a lecturer at the University of South Wales he is now Director of the Brecon Beacons Observatory, a Dark Sky Ambassador for the Brecon Beacons National Park, a committee member of the International Dark Sky Reserve Steering Groups of Snowdonia and Brecon Beacons National Parks, the Eryri Dark Sky Partnership and a science presenter and astronomer at Dark Sky Wales.

He is a consultant to the Welsh Government through his involvement with the Dark Sky Discovery initiative, enabling public access to dark sky sites in association with Dark Sky Wales, Dark Sky Scotland and Natural England. He was also responsible for surveying the sky quality of the Brecon Beacons National Park for their successful bid to gain International Dark Sky Association Dark Sky Reserve status in 2013. Martin has spent many years directing stargazing events nationally and in the last three years has provided training for the BBNPA dark sky ambassadors in dark sky related business matters.

Martin is a Fellow of the Royal Astronomical Society; A Fellow of the Higher Education Academy; a member of the Astrobiology Society of Britain; the European Society for the History of Science; the British Astronomical Association; the British Science Association; the Webb Deep-Sky Society; the Society for Popular Astronomy and the Astronomical League. He is also a local representative for the BAA Commission for Dark Skies.